STAIR LAYOUT

Stanley Badzinski, Jr.

AMERICAN TECHNICAL PUBLISHERS, INC.
HOMEWOOD, ILLINOIS 60430

© 1996 by American Technical Publishers, Inc.
All rights reserved

1 2 3 4 5 6 7 8 9 – 96 – 9 8 7 6 5 4 3 2

Printed in the United States of America

ISBN 0-8269-0701-6

CONTENTS

1	Stair Basics	1
	Trade Test 1	17
2	Stair Design	21
	Trade Test 2	33
3	Stairwells	37
	Trade Test 3	45
4	Headroom	51
	Trade Test 4	61
5	Landings	67
	Trade Test 5	85
6	Stringers	91
	Trade Test 6	119
7	Winders	123
	Trade Test 7	135
8	Stair Safety	139
	Final Exam	143
	Answers	149
	Appendix	155
	Glossary	161
	Index	167

INTRODUCTION

Stair Layout contains the basic principles of stair layout practices. Stair terminology, print information, stairway design, building codes, and stairway types are discussed.

The three stair ratio formulas in general use are introduced. Examples of each formula are given. Stairwells are presented with examples of determining stairwell length. Headroom procedures are detailed. Landings in straight, quarter-turn, and half-turn stairs are discussed.

Step-by-step procedures show how to lay out stringers. The story pole and framing square are the primary tools utilized. Winders are designed using basic formulas. Safety is emphasized in stair design.

Chapters 1 through 7 are followed by Trade Tests. A Final Exam follows Chapter 8. Answers for Trade Tests 1 through 7 and the Final Exam are given in the back of this book. The Glossary and Appendix contains illustrated terms and useful tables.

The Publisher

1 STAIR BASICS

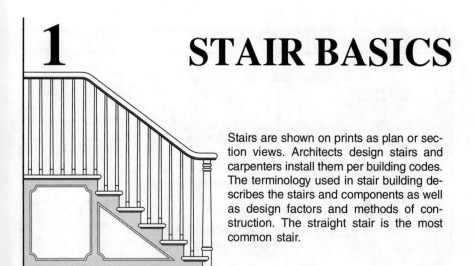

Stairs are shown on prints as plan or section views. Architects design stairs and carpenters install them per building codes. The terminology used in stair building describes the stairs and components as well as design factors and methods of construction. The straight stair is the most common stair.

PLANS

The architect designs stairs that comply with building codes and fits the particular space of the building. Many times the carpenter is called upon to lay out and build stairs. The carpenter who installs the stairs has the responsibility of working out details of construction not covered specifically on the prints. Nearly as often, the cabinetmaker or millworker is called upon to measure a building, and lay out and build stairs in the mill. These stairs are installed by the carpenter on the job.

All stairs are shown on prints as conventional representations. A *conventional representation* is a simplified way of representing building components on a print. Threaded fasteners (such as screws, bolts, and nuts) are often drawn as conventional representations to save drawing time and produce a consistent appearance.

Stairs may be shown in either plan view or sectional view. A *plan view* is a view looking down on the object. A *sectional view* is a view made by passing a cutting plane through the object. See Figure 1-1.

2 STAIR LAYOUT

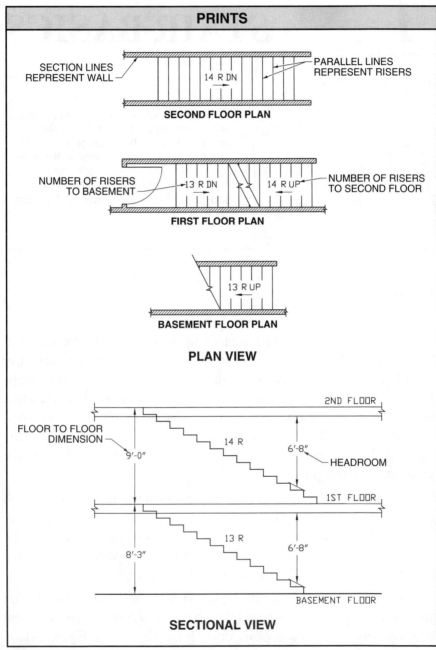

Figure 1-1. Stairs are shown as plan or sectional views on prints.

Dimensions on plans provide critical information for the stair builder. Typical dimensions include floor to floor dimensions and the headroom dimension.

Dimensions on plans are given with a dash separating the foot and inch numerals to promote clarity. For example, a dimension that is nine feet long is given as 9'-0". A dimension that is six feet, eight inches long is given as 6'-8". Additionally, dimensions may be given in decimal format. For example, the decimal equivalent of $80\frac{1}{2}"$ is 80.50". The number of places to the right of the decimal point indicate the preciseness of the dimension. A dimension of .50" is measured in $\frac{1}{100}$ ths of an inch. A dimension of .5" is measured in $\frac{1}{10}$ ths of an inch.

All dimensions are terminated by arrowheads showing the exact starting and ending points for that particular dimension. Dimensions are carefully lettered to avoid confusion.

The conventional representation for stairs on plan views is a series of parallel lines representing risers. Designations such as *14 R UP* (14 risers up) and *13 R DN* (13 risers down) indicate the number of risers to the floor immediately above or below. An arrow with the designation points up or down, depending upon the direction of travel.

The conventional representation for showing stairs in section views is as a cross section of the stairs in elevation. This view clearly shows the floors above and below with the floor-to-floor dimension between them given. The number of risers and the headroom is also given.

STAIR TERMINOLOGY

The terminology used in stair building describes the stairs and components as well as design factors and methods of construction. Common terms used in stair building include the following. See Figure 1-2. See Glossary.

4 STAIR LAYOUT

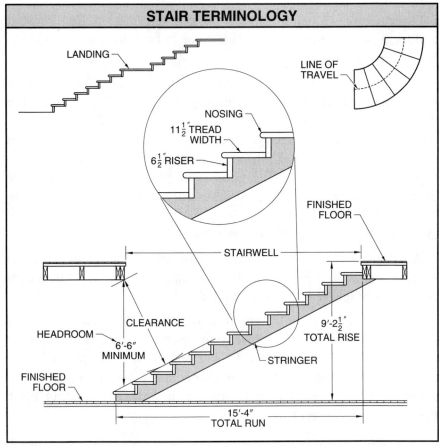

Figure 1-2. Terminology used in stair building describes the stairs.

Headroom. The vertical distance from the floor construction above the stair (at the end of the stairwell) to the slope of the stair.

Landing. A horizontal platform separating two flights of stairs.

Line of travel. The line along which most people walk on a stairway.

Nosing. The portion of the tread which projects beyond the riser face.

Riser. The vertical board between treads.

Stair ratio. The formula expressing the relationship between unit rise and unit run.

Story pole. A strip of lumber (usually a 1 × 2 or 2 × 2) on which the total rise of the stair is marked.

Stringer (carriage). The part of the stair construction which is cut out to receive the risers and treads. The stringer (carriage) supports the steps.

Total rise. The total vertical distance (height) of the stairs. It is measured from finished floor to finished floor.

Total number of rises (risers). The number of unit rises (risers) in the stairway.

Total number of treads. The number of unit runs in a given stairway. It is always one less than the total number of rises.

Total run. The total horizontal distance (length) of the stairs.

Tread. The horizontal part of the stair which is walked on.

Tread width. The width of the tread plus the width of the nosing.

Unit rise. The height of each riser.

Unit run. The width of each tread, not including the nosing.

Upper construction. The total thickness of the floor and ceiling construction over the stairway.

Wellhole. The space or opening in the floor through which the stair passes.

STAIRWAY DESIGN

In the design of any stairway, ease of ascending and descending is of prime importance. If the treads are too wide and the rise too small, the stair will be uncomfortable to use. If the stairway is too steep, with a narrow tread and high riser, any person ascending the stair will be subjected to an unnecessary strain.

To avoid either condition, an acceptable stair ratio formula should be employed. Stair ratio is the ratio between the unit rise and unit tread of a set of stairs expressed as a formula ($T + R = 17''$ to $18''$). A rule-of-thumb formula for determining the stair ratio is *the width of the tread*

6 STAIR LAYOUT

plus the height of the riser shall be 17″ to 18″. Applying this rule-of-thumb, some common riser and tread dimensions are:

RISER	TREAD	TOTAL
6″	11″	17″
6½″	11½″	18″
6¾″	11¼″	18″

Many other riser and tread dimensions are possible. Architects design, and carpenters build, stairs with risers and treads to fit within accepted dimensions and meet various codes.

Stair pitch is the slope (angle) of a set of stairs. Preferred angles for stairs are between 30° and 35°. However, stairs with a pitch of 40° are suitable if the ratio between rise and run is maintained. See Figure 1-3. Lesser angles produce larger total runs. Higher angles produce stairways that are difficult to climb and may be too steep.

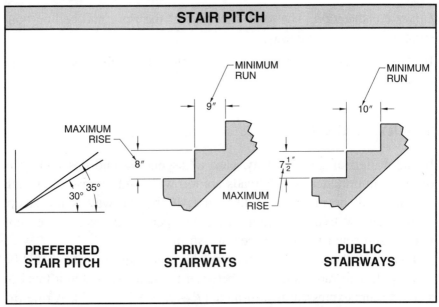

Figure 1-3. Stair pitch is the slope (angle) of a set of stairs.

BUILDING CODES

The importance of properly designed, properly built, and properly installed stairs cannot be over emphasized. Every year many injuries and deaths are attributed to falls on poorly designed and poorly built stairs.

To help prevent accidents on stairways, various building codes have established requirements for different types of stairways. Building codes vary, but all give maximum riser height and minimum tread width.

Any stairway that has two or more risers must comply to minimum code standards. All stairways serving an occupancy load of 49 people or less must be at least 36" wide. All stairways serving an occupancy load of 50 people or more must be at least 44" wide. See Figure 1-4.

The unit rise and unit run should remain constant from bottom to top of a flight of stairs. Riser height variance in a flight of stairs shall not exceed $\frac{3}{8}"$. Good building practice does not exceed $\frac{1}{8}"$ of riser height variance.

Stairways for dwellings and stairways less than 44" wide must have at least one handrail. Stairways 44" or more in width must have at least two handrails. An intermediate handrail is required for stairways more than 88" wide. See Figure 1-5.

Handrails are located 30"–36" above the nosing of the stairway treads and must run continuously for the full length of the stairway. Handrails in commercial buildings are required to extend 6" beyond the top and bottom riser. These handrails may be required to return to a wall or be terminated in newel posts. The cross-sectional dimension of a handrail must be not less than $1\frac{1}{4}"$ nor more than 2" and must have a smooth surface with no sharp corners. Handrails must be provided with an opening space of not less than $1\frac{1}{2}"$ between wall and handrail.

8 STAIR LAYOUT

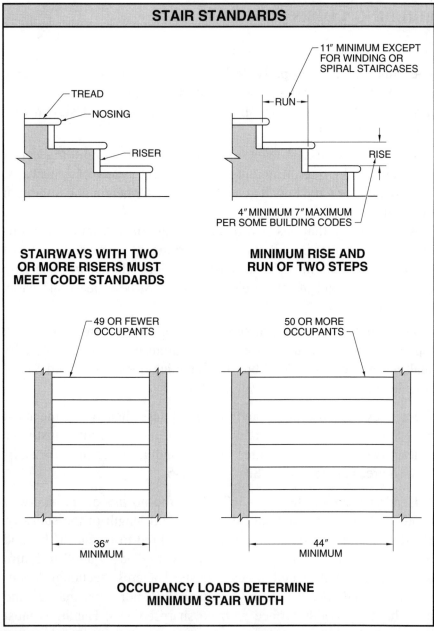

Figure 1-4. Any stairway that has two or more risers must comply to minimum code standards.

Stair Basics 9

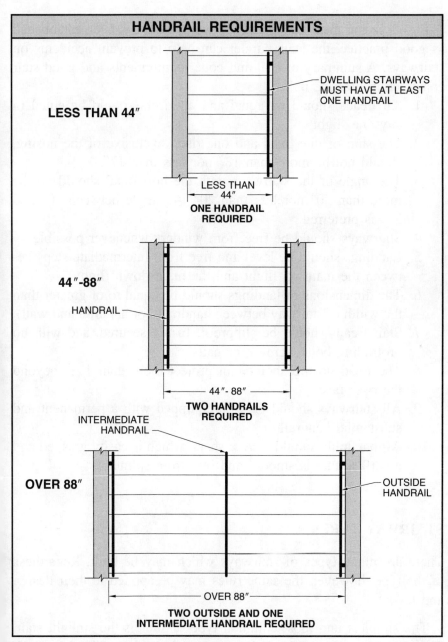

Figure 1-5. Stairways for dwellings and stairways less than 44″ wide must have at least one handrail.

10 STAIR LAYOUT

By observing these codes and following what has been established as good practice, the stair builder can help to prevent accidents on stairways. A summary of building code requirements and good stair construction practice is:
1. All treads should be equal and all risers should be equal in any one flight.
2. The sum of one tread and one riser, exclusive of the nosing, should not be more than 18″ nor less than 17″.
3. The angle of the stairways with the horizontal should not be more than 50° nor less than 20°. An angle between 30° and 35° is preferred.
4. Stairways should be free from winders whenever possible.
5. Landings should be level and free from intermediate steps between the main up flight and the main down flight.
6. The dimensions of landings should be equal to or greater than the width of stairway between handrails (or handrail and wall).
7. Stair treads should be slipproof, firmly secured, and with no protruding bolts, screws, or nails.
8. The tread nosing should not project more than 1¾″ beyond the riser face.
9. All stairways should be well equipped with a permanent and substantial handrail(s).
10. All handrails should have a shape which is easily grasped and a surface that is smooth and free from splinters.

STAIRWAY TYPES

There are many types of stairways which may be built. Regardless of the type, however, the same rules may be applied to their design and layout.

The simplest and most common type of stair is the straight stair. The straight stair may be built in a number of combinations with landings installed where the stair changes directions. See Figure 1-6.

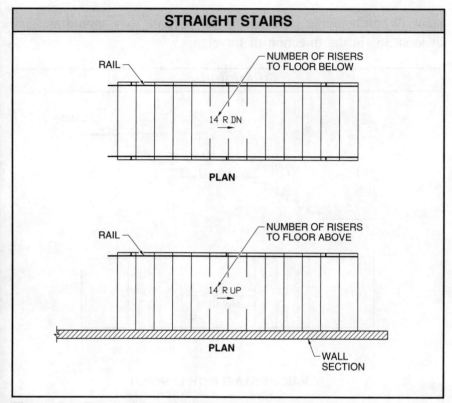

Figure 1-6. The straight stair is the most common type of stair.

Landings for stairways must extend in the direction of travel equal to the stairway width. A landing is a platform to change the direction of a stairway or to break the run. For straight runs, the distance need not exceed 44". The minimum length of a landing should be 36". Distances between landings shall not exceed 12'-0" vertically. Two of the most common uses of landings are in the quarter-turn stair and the half-turn stair. See Figure 1-7.

Landings are also used to break a long straight stairway into two or more flights of straight stairs. Landings in this type of stair are desirable as a resting place for those climbing stairs and they also serve to stop the fall of anyone unfortunate enough to slip on the

12 STAIR LAYOUT

steps. When landings are used for this purpose they should measure at least 36″ in the direction of travel.

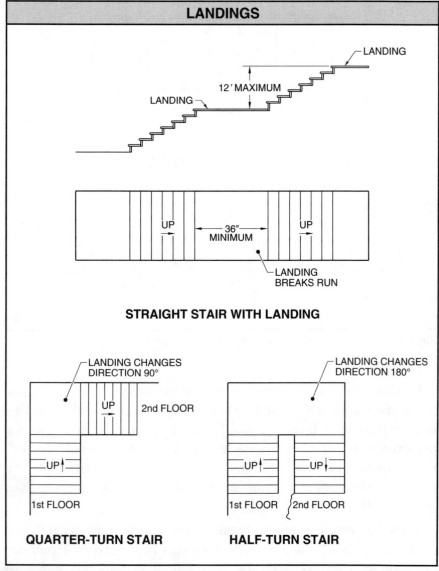

Figure 1-7. A landing is a platform to change the direction of a stairway or to break the run.

Winder (winding) stairs should be avoided if possible as more accidents are likely to occur on poorly designed winders than on straight stairs. However, if winding treads must be used because of space limitations, winder layouts following good design results in a reasonably safe stair. Two of the most popular winder arrangements are the quarter-turn and the half-turn. See Figure 1-8.

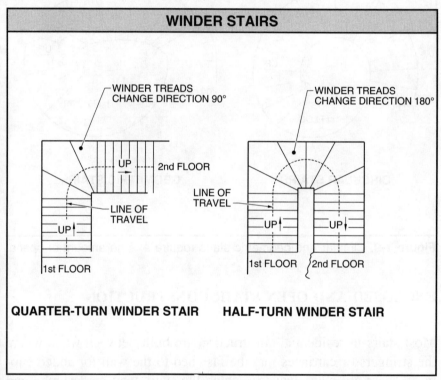

Figure 1-8. Winder stairs change direction with winding treads.

In addition to the stair types mentioned there are many possible arrangements of geometric and circular stairs. These require a considerable amount of space and are installed only in more expensive houses and in some commercial buildings. The stair design rules which apply to straight and winding stairs may be applied to circular and geometric stairs. See Figure 1-9.

14 STAIR LAYOUT

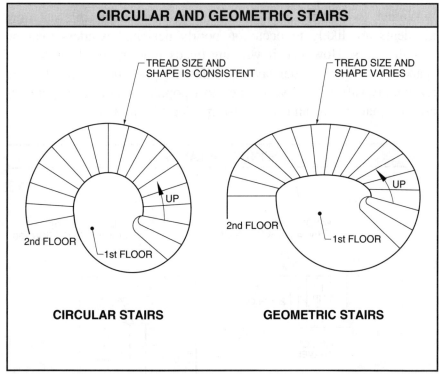

Figure 1-9. Circular and geometric stairs require a large amount of space.

ENCLOSED AND OPEN STAIR CONSTRUCTION

Most stairs in residential construction are built between walls where the stringers or carriages may be fastened to the wall for added support. There are many ways in which the stairs may be designed and built. In a half-open stringer stair, two to five treads are typically exposed on one side. The balustrade may end at the wall or it may bypass the wall. See Figure 1-10.

With an open stringer stair, one side of the stairs is in the open and can start with a bullnose step and turnout, or it may start with a regular step and a newel post. This is sometimes referred to as an enclosed stairway. It is economical to build because of the lack of a balustrade. A *balustrade* is a row of balusters topped by a rail. A *baluster* is a vertical support for a rail.

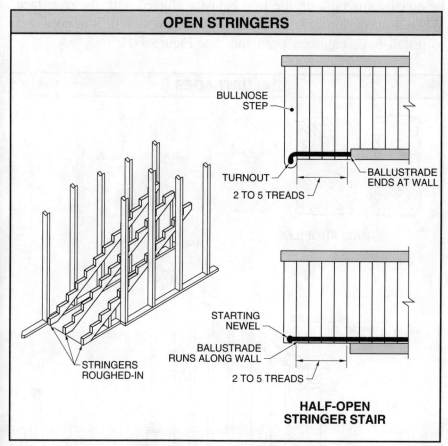

Figure 1-10. Open stringer stairs are supported by the wall on one side.

The stairs are exposed on one side or both sides in open stair construction. This makes the stairs a prominent feature of the building and adds to the cost of construction due to the need for self-supporting stringers and an extensive balustrade.

Many balustrade combinations are possible for an open stairway. However, in all cases it is necessary to place the rail (banister) at the minimum height required by the code and to space the balusters

16 STAIR LAYOUT

equally. If the code allows a maximum baluster spacing of 6″, it is usually possible to have two balusters per tread. The balusters must be equally spaced with the first baluster aligned with the riser face. When a maximum baluster spacing of 4″ is required, it is necessary to install three balusters per tread. See Figure 1-11.

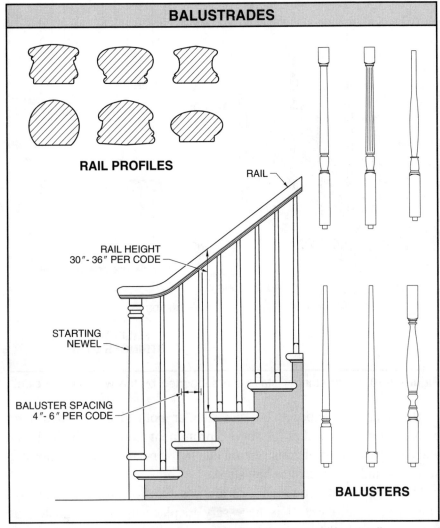

Figure 1-11. A balustrade is a row of balusters topped by a rail.

TRADE TEST 1
STAIR BASICS

Date _____ Name _____

_____ 1. All stairs are shown on prints as _____ representations.

    A. conventional   C. institutional
    B. commercial   D. neither A, B, nor C

_____ 2. The _____ is the portion of the tread which projects beyond the riser face.

    A. overhang   C. bullnose
    B. nosing   D. neither A, B, nor C

_____ 3. Stair pitch is the _____ or angle of a set of stairs.

_____ 4. The minimum rise of a step is _____".

_____ 5. Riser height variance in a flight of stairs shall not exceed _____".

_____ 6. The open space between the wall and handrail must be at least _____".

_____ 7. A(n) _____ view is a view looking down on the object.

_____ 8. A(n) _____ is a view made by passing a cutting plane through the object.

18 STAIR LAYOUT

_____ 9. The _____ is the line along which most people walk on a stairway.

_____ 10. The _____ is the horizontal part of the stair which is walked on.

_____ 11. The _____ is the space or opening in the floor through which the stair passes.

        A. landing   C. cutout
        B. wellhole   D. neither A, B, nor C

_____ 12. The minimum preferred stair pitch at A is _____°.

_____ 13. The maximum preferred stair pitch at B is _____°.

_____ 14. The maximum rise at C is _____".

_____ 15. The minimum run at D is _____".

_____ 16. The maximum rise at E is _____".

_____ 17. The minimum run at F is _____".

PRIVATE STAIRWAYS

PUBLIC STAIRWAYS

_____ 18. The total _____ is the total horizontal distance (length) of the stairs.

_____ 19. The vertical board between treads is the _____.

_____ 20. A(n) _____ is a strip of lumber on which the total rise of the stair is marked.

_____ 21. The _____ rise is the height of each riser.

_____ 22. The _____ rise is the total vertical distance (height) of the stairs.

_____ 23. All stairways serving an occupancy load of _____ people or less must be at least 36″ wide.
A. 24   C. 76
B. 49   D. 101

_____ 24. The cross-sectional dimension of a handrail must be no less than _____″.
A. ¾   C. 1¼
B. 1   D. 1½

_____ 25. The cross-sectional dimension of a handrail must be no more than _____″.
A. 1   C. 1½
B. 1¼   D. 2

T   F   26. The sum of one tread and one riser should not exceed 18″.

T   F   27. The maximum angle of the stairway with the horizontal should not exceed 40°.

20 STAIR LAYOUT

_____ **28.** The tread nosing should not project more than _____" beyond the riser face.

_____ **29.** One handrail is required when G is less than _____".

_____ **30.** Two handrails are required when H is from _____" to _____".

|LESS THAN "| |← "- "→| |← OVER "→|
Ⓖ Ⓗ Ⓘ

_____ **31.** Two outside and one intermediate handrails are required when I is over _____".

_____ **32.** The simplest and most common type of stair is the _____ stair.

2 STAIR DESIGN

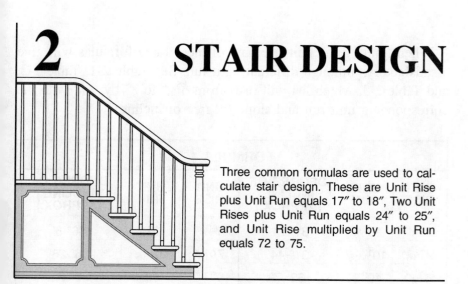

Three common formulas are used to calculate stair design. These are Unit Rise plus Unit Run equals 17″ to 18″, Two Unit Rises plus Unit Run equals 24″ to 25″, and Unit Rise multiplied by Unit Run equals 72 to 75.

STAIR DESIGN

The initial design of a stairway should be made without regard to wellhole size or other space limitations, but consideration should be given to the slope or angle of incline of the stair. A stair with a lesser incline is easier to ascend than one with a steep incline, provided that the ratio between the unit rise and unit run is acceptable. Therefore, layout should be made with safety precautions and ease of use as prime considerations. The final design must also meet local building code requirements.

As a general rule, the unit rise of a stair should be kept between 6⅝″ and 7¾″. However, basement stairways with a unit rise of 8″ are satisfactory if the unit run is determined by using one of the accepted stair ratio formulas.

Three stair ratio formulas in general use are:

Formula 1. *Unit Rise plus Unit Run equals 17″ to 18″.*
Formula 2. *Two Unit Rises plus Unit Run equals 24″ to 25″.*
Formula 3. *Unit Rise multiplied by Unit Run equals 72 to 75.*

22 STAIR LAYOUT

When limiting the unit rise, any of these three formulas will give a satisfactory, although different, unit run. See Table 2-1, Table 2-2, and Table 2-3, which list unit rises from $6\frac{5}{8}''$ to $8''$ by $\frac{1}{16}''$ with the corresponding unit run and slope (degree of incline).

FORMULA 1					
UNIT RISE	UNIT RUN	DEGREE OF INCLINE (APPROX.)	UNIT RISE	UNIT RUN	DEGREE OF INCLINE (APPROX.)
$6\frac{5}{8}''$	$10\frac{7}{8}''$	31°-21'	$7\frac{3}{8}''$	$10\frac{1}{8}''$	36°-5'
$6\frac{11}{16}''$	$10\frac{13}{16}''$	31°-44'	$7\frac{7}{16}''$	$10\frac{1}{16}''$	36°-28'
$6\frac{3}{4}''$	$10\frac{3}{4}''$	32°-7'	$7\frac{1}{2}''$	$10''$	36°-52'
$6\frac{13}{16}''$	$10\frac{11}{16}''$	32°-31'	$7\frac{9}{16}''$	$9\frac{15}{16}''$	37°-16'
$6\frac{7}{8}''$	$10\frac{5}{8}''$	32°-54'	$7\frac{5}{8}''$	$9\frac{7}{8}''$	37°-40'
$6\frac{15}{16}''$	$10\frac{9}{16}''$	33°-18'	$7\frac{11}{16}''$	$9\frac{13}{16}''$	38°-5'
$7''$	$10\frac{1}{2}''$	33°-40'	$7\frac{3}{4}''$	$9\frac{3}{4}''$	38°-29'
$7\frac{1}{16}''$	$10\frac{7}{16}''$	34°-5'	$7\frac{13}{16}''$	$9\frac{11}{16}''$	38°-53'
$7\frac{1}{8}''$	$10\frac{3}{8}''$	34°-29'	$7\frac{7}{8}''$	$9\frac{5}{8}''$	39°-17'
$7\frac{3}{16}''$	$10\frac{5}{16}''$	34°-52'	$7\frac{15}{16}''$	$9\frac{9}{16}''$	39°-41'
$7\frac{1}{4}''$	$10\frac{1}{4}''$	35°-16'	$8''$	$9\frac{1}{2}''$	40°-5'
$7\frac{5}{16}''$	$10\frac{3}{16}''$	35°-40'			

Table 2-1. Unit rise plus unit run equals 17" to 18".

These tables were developed as a means for comparing the three stair ratio formulas and not necessarily as a guide for designing a stair as it is very doubtful that the unit rise will work out to an even $\frac{1}{16}''$. Because more than one solution is possible with the three stair ratio formulas, the formulas were re-written using the average in each formula. In these three tables:

Table 2-1. Unit Run equals 17½″ − Unit Rise.
Table 2-2. Unit Run equals 24½″ − Two Unit Rises.
Table 2-3. Unit Run equals 73½ ÷ Unit Rise.

FORMULA 2					
UNIT RISE	UNIT RUN	DEGREE OF INCLINE (APPROX.)	UNIT RISE	UNIT RUN	DEGREE OF INCLINE (APPROX.)
6⅝″	11¼″	30°-29′	7⅜″	9¾″	37°-6′
6¹¹⁄₁₆″	11⅛″	31°	7⁷⁄₁₆″	9⅝″	37°-42′
6¾″	11″	31°-32′	7½″	9½″	38°-17′
6¹³⁄₁₆″	10⅞″	32°-4′	7⁹⁄₁₆″	9⅜″	38°-52′
6⅞″	10¾″	32°-36′	7⅝″	9¼″	39°-30′
6¹⁵⁄₁₆″	10⅝″	33°-8′	7¹¹⁄₁₆″	9⅛″	40°-7′
7″	10½″	33°-40′	7¾″	9″	40°-44′
7¹⁄₁₆″	10⅜″	34°-12′	7¹³⁄₁₆″	8⅞″	41°-21′
7⅛″	10¼″	34°-48′	7⅞″	8¾″	42°
7³⁄₁₆″	10⅛″	35°-21′	7¹⁵⁄₁₆″	8⅝″	42°-38′
7¼″	10″	35°-56′	8″	8½″	43°-15′
7⁵⁄₁₆″	9⅞″	36°-31′			

Table 2-2. Two unit rises plus unit run equals 24″ to 25″.

Formula 1 is the most simple and, therefore, the easiest to work with. It gives a stair that is slightly steeper than the other formulas when unit rises are below 7″. However, at unit rises over 7″, it gives a stair with a lesser incline than Formulas 2 and 3. Therefore, Formula 1 is the most desirable to use when the unit rise is over 7″ and the angle of incline is to be held to a minimum.

STAIR LAYOUT

FORMULA 3

UNIT RISE	UNIT RUN	DEGREE OF INCLINE (APPROX.)	UNIT RISE	UNIT RUN	DEGREE OF INCLINE (APPROX.)
6⅝"	11.09"	30°-51'	7⅜"	9.96"	36°-31'
6¹¹⁄₁₆"	10.99"	31°-19'	7⁷⁄₁₆"	9.88"	36°-58'
6¾"	10.89"	31°-47'	7½"	9.80"	37°-26'
6¹³⁄₁₆"	10.79"	32°-16'	7⁹⁄₁₆"	9.72"	37°-53'
6⅞"	10.69"	32°-45'	7⅝"	9.64"	38°-21'
6¹⁵⁄₁₆"	10.59"	33°-14'	7¹¹⁄₁₆"	9.56"	38°-49'
7"	10.50"	33°-40'	7¾"	9.48"	39°-16'
7¹⁄₁₆"	10.41"	34°-8'	7¹³⁄₁₆"	9.41"	39°-43'
7⅛"	10.32"	34°-37'	7⅞"	9.33"	40°-10'
7³⁄₁₆"	10.23"	35°-6'	7¹⁵⁄₁₆"	9.25"	40°-36'
7¼"	10.14"	35°-34'	8"	9.19"	41°-2'
7⁵⁄₁₆"	10.05"	36°-13'			

Table 2-3. Unit rise multiplied by unit run equals 72 to 75.

Formulas 2 and 3 are desirable when the unit rise is below 7" and the angle of incline is to be held to a minimum. These two formulas are also desirable when a lack of space requires the installation of a stair with a unit rise of over 7" and slightly greater incline.

The basic order in designing stairs is determining:
1. Total Rise
2. Number of Risers
3. Unit Rise
4. Unit Run
5. Total Run

Total Rise

Total rise is the vertical distance from finished floor to finished floor. The total rise may be determined from an elevation sectional view of the building plans.

Number of Risers

To find the number of risers in a flight of stairs, divide the total rise in inches by 7" (Method A). The result is usually a whole number and a remainder. The whole number is the number of risers in the stair. The remainder is disregarded at this point. Using Method A to determine the number of risers in a stair generally results in a greater number of risers with a smaller unit rise and a lesser incline.

For example, what is the number of risers in a flight of stairs with a total rise of $9'\text{-}10\frac{1}{2}'''$?

Method A
Total Rise = $9'\text{-}10\frac{1}{2}''$ = $118\frac{1}{2}''$
$118.5'' \div 7'' = 16+$
Number of Risers = **16**

A second method of determining the number of risers in a flight of stairs is to divide the total rise in inches by 8" (Method B). The result is usually a whole number and a remainder. The remainder is rounded off to the next full number to find the number of risers. Using Method B to determine the number of risers in a stair generally results in a smaller number of risers for a given total rise but with a greater unit rise and greater incline.

Method B
Total Rise = $9'\text{-}10\frac{1}{2}''$ = $118\frac{1}{2}''$
$118.5'' \div 8'' = 14+$
Number of Risers = **15**

26 STAIR LAYOUT

The stair with the larger number of risers in a given total rise has a smaller incline and, therefore, is easier to ascend. However, the stair with fewer risers in a given total rise occupies less horizontal space. These facts may influence the choice of one method over the other in determining the number of risers in a stairway. See Figure 2-1.

Unit Rise

To find the unit rise in a flight of stairs, divide the total rise by the number of risers. The answer should be carried to the nearest .01″.

For example, what is the unit rise in a flight of stairs with a total rise of 9′-10½″ and 16 risers?

Total Rise = 9′-10½″ = 118½″
Number of Risers = 16
118.5″ ÷ 16 = 7.406″
Unit Rise = **7.41″**

For example, what is the unit rise in a flight of stairs with a total rise of 9′-10½″ and 15 risers?

Total Rise = 9′-10½″ = 118½″
Number of Risers = 15
118.5″ ÷ 15 = 7.9″
Unit Rise = **7.9″**

Unit Run

The unit run of a stair is determined after the unit rise for the stair has been established. To calculate the unit run, the designer must first decide which formula to use. It is generally most productive to begin with Formula 1, *Unit Rise plus Unit Run equals 17″ to 18″*. If any adjustment in the unit run is necessary, the other formulas may be used.

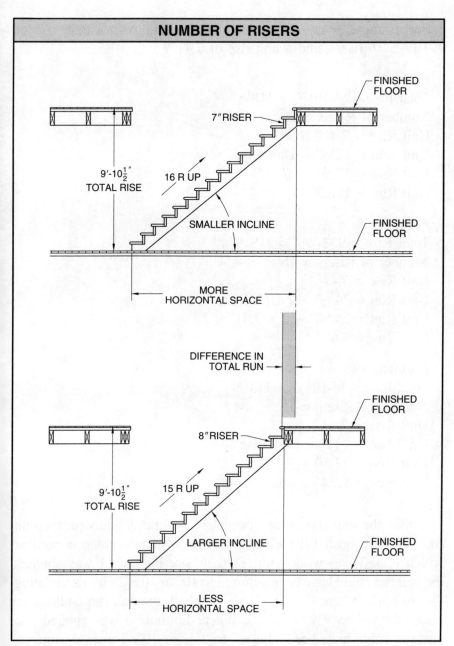

Figure 2-1. The stair with the larger number of risers in a given total rise has a smaller incline and occupies more horizontal space.

28 STAIR LAYOUT

For example, what is the unit run of a stair with a total rise of 9'-10½", 16 risers, and a unit rise of 7.41"?

Formula 1
Total Rise = 9'-10½" = 118½"
Number of Risers = 16
Unit Rise = 7.41"
Unit Run = 17½" − Unit Rise
Unit Run = 17.50" − 7.41"
Unit Run = **10.09"**

Formula 2
Total Rise = 9'-10½" = 118½"
Number of Risers = 16
Unit Rise = 7.41"
Unit Run = 24½" − 2 Unit Rises
Unit Run = 24.50" − 2 × 7.41"
Unit Run = **9.68"**

Formula 3
Total Rise = 9'-10½" = 118½"
Number of Risers = 16
Unit Rise = 7.41"
Unit Run = 73½" ÷ Unit Rise
Unit Run = 73.50 ÷ 7.41"
Unit Run = **9.92"**

Unlike the unit rise, which cannot be arbitrarily adjusted, the unit run may be adjusted if for no other reason than to obtain a number which is easier to work with. Thus, where Formula 1 was applied, the unit run could have been adjusted to 10" or 10⅛" without violating the formula. Where Formula 2 was applied, the unit run could have been adjusted to 9⅝" or 9¾". Where Formula 3 was applied, the unit run could have been adjusted to 9⅞" or 10". These adjustments would not have violated the respective formulas.

Total Run

The total run of a stairway is determined by multiplying the unit run by the number of unit runs in the stairway. Since there is always one less tread than there are risers in a given flight of stairs, there is one less unit run than there are unit rises. See Figure 2-2.

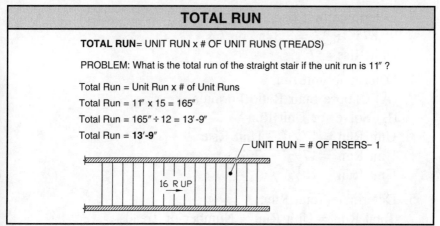

Figure 2-2. The total run is found by multiplying the unit run times the number of unit runs (always 1 less than the number of risers).

For example, what is the total run of a stairway with 16 risers and a unit run of 10″?

16 Risers
15 Treads or Unit Runs
Unit Run = 10″
Total Run = 15 × 10″ = 150″
150″ ÷ 12″ = 12′-6″
Total Run = **12′-6″**

Stair Design Problem

A print shows a total height of $95\frac{7}{8}″$ between the first and second floor of a dwelling. Find the total rise, number of risers, unit rise, unit run, and total run.

1. Determine total rise.
   Total Rise = **95⅞″**

2. Determine number of risers.
   95.875″ ÷ 7 = 13
   Risers = **13**

3. Determine unit rise.
   95.875″ ÷ 13 = 7.375″ or 7⅜″
   Unit Rise = **7⅜″**

4. Determine unit run.
   (A) Choose Stair Ratio Formula
   (B) Solve for Unit Run
   Unit Run = 17½″ − Unit Rise
   Unit Run = 17½″ − 7⅜″
   Unit Run = **10⅛″**

5. Determine Total Run.
   Total Run = Unit Run × Number of Treads
   Total Run = 10⅛″ × 12 = 121½″
   121½″ ÷ 12 = 10′-1½″
   Total Run = **10′-1½″**

Existing Conditions

Occasionally a stair designer or stair builder is called upon to build a stair where both the total rise and total run are fixed. This condition may arise when a stair must be placed between a floor and a landing which is already in place. When confronted with a problem of this type, the initial procedure is the same; that is, determine the number of risers first and then the unit rise.

A change in procedure is made when determining unit run. The unit run is found by dividing the total available run by the number of treads, and the result is checked to see if the unit rise and unit run fit into the stair ratio formula being used. If the results are unsatisfactory, a second solution should be attempted.

For example, what is the number of risers, unit rise, and unit run of a flight of stairs with a total rise of $85\frac{1}{2}''$ and a total run of 99.0''?

Solution A
Number of Risers
$85\frac{1}{2}'' \div 7'' = 12+$ or **12**

Unit Rise
$85\frac{1}{2}'' \div 12 = \mathbf{7\frac{1}{8}''}$

Unit Run = Total Run ÷ Number of Treads
$99'' \div 11 = \mathbf{9''}$

Check: Unit Rise + Unit Run = 17'' to 18''
$7\frac{1}{8}'' + 9'' = 16\frac{1}{8}''$
Unit Rise + Unit Run = $16\frac{1}{8}''$ (This solution is not acceptable as it does not meet the requirements of the stair ratio formula).

Solution B
Number of Risers
$85\frac{1}{2}'' \div 8'' = 10+$ or **11**

Unit Rise
$85\frac{1}{2}'' \div 11 = \mathbf{7.77''}$

Unit Run = Total Run ÷ Number of Treads
$99'' \div 10 = \mathbf{9.9''}$

Check: Unit Rise + Unit Run = 17'' to 18''
$7.77'' + 9.9'' = 17.67''$
Unit Rise + Unit Run = $17.67''$ (This solution is acceptable as it does meet the requirements of the stair ratio formula).

The preceding examples point out that there is more than one solution possible to a stair design problem. However, a logical solution is not easily obtained in all cases when total rise and total run are fixed. The stair builder must compromise with the formulas and pick a solution which approximates the stair formulas.

32 STAIR LAYOUT

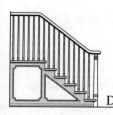

TRADE TEST 2
STAIR DESIGN

Date _____ Name _____

_____ 1. As a general rule, the unit rise of a stair should be kept between _____" and _____".

A. 6⅛; 7⅛   C. 6½; 7½
B. 6¼; 7⅜   D. 6⅝; 7¾

T   F   2. Stair layout should be made with safety precautions and ease of use as prime considerations.

T   F   3. Generally, basement stairways with a unit rise of 8" are satisfactory.

_____ 4. _____ is the vertical distance from finished floor to finished floor.

_____ 5. The unit run for a stair is determined after the unit _____ for the stair has been established.

_____ 6. The total run of a stairway is determined by multiplying the _____ times the number of unit runs in the stairway.

_____ 7. Formula 1 states *Unit Rise plus Unit Run equals* _____" to _____".

A. 16; 17   C. 17; 18
B. 16½; 17   D. neither A, B, nor C

34 STAIR LAYOUT

_____ 8. Formula 2 states *Two Unit Rises plus Unit Run equals* _____ " *to* _____ ".

    A. 23; 24   C. 23; 25
    B. 24; 25   D. neither A, B, nor C

_____ 9. Formula 3 states *Unit Rise multiplied by Unit Run equals* _____ *to* _____ .

    A. 72; 75   C. 74; 80
    B. 72; 76   D. 74; 82

_____ 10. Total rise may be determined from an elevation _____ view of the building plans.

Ⓐ Ⓑ Ⓒ
$96\frac{1}{2}$"   108" $110\frac{1}{2}$"

Use Formula 1 for Problems 11 through 19.

_____ 11. The number of risers at A is _____ .

_____ 12. The unit rise at A is _____ ".

_____ 13. The unit run at A is _____ ".

_____ 14. The number of risers at B is _____.

_____ 15. The unit rise at B is _____".

_____ 16. The unit run at B is _____".

_____ 17. The number of risers at C is _____.

_____ 18. The unit rise at C is _____".

_____ 19. The unit run at C is _____".

Use Formula 2 for Problems 20 through 28.

_____ 20. The number of risers at A is _____.

_____ 21. The unit rise at A is _____".

_____ 22. The unit run at A is _____".

_____ 23. The number of risers at B is _____.

_____ 24. The unit rise at B is _____".

_____ 25. The unit run at B is _____".

_____ 26. The number of risers at C is _____.

_____ 27. The unit rise at C is _____".

_____ 28. The unit run at C is _____".

36 STAIR LAYOUT

Use Formula 3 for Problems 29 through 37.

_____ **29.** The number of risers at A is _____.

_____ **30.** The unit rise at A is _____″.

_____ **31.** The unit run at A is _____″.

_____ **32.** The number of risers at B is _____.

_____ **33.** The unit rise at B is _____″.

_____ **34.** The unit run at B is _____″.

_____ **35.** The number of risers at C is _____.

_____ **36.** The unit rise at C is _____″.

_____ **37.** The unit run at C is _____″.

3 STAIRWELLS

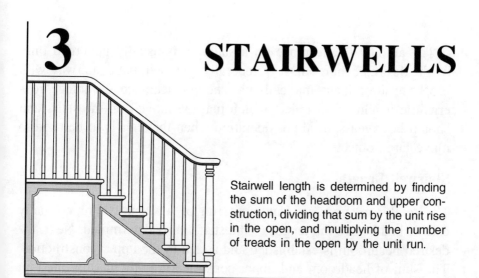

Stairwell length is determined by finding the sum of the headroom and upper construction, dividing that sum by the unit rise in the open, and multiplying the number of treads in the open by the unit run.

STAIRWELL CALCULATION

A stairway should be constructed in a manner which will allow a person to descend without having to stoop or bend over. To obtain proper headroom, the stairwell must be made long enough to uncover, or leave in the open, a sufficient number of risers and treads. A headroom distance of 6'-4" is acceptable, under most building codes, for basement stairways in residential work. However, it is desirable to maintain greater headroom whenever possible.

Main stairways in residential work should have a minimum headroom of 6'-8" and, whenever possible, the minimum headroom in a dwelling should be increased to 7'-0" or more. Stairways with adequate headroom make it possible for people to walk up or down without bending over to avoid bumping their heads on the upper construction. Stairways with adequate headroom also allow sufficient vertical space for moving large items such as furniture, refrigerators, etc.

38 STAIR LAYOUT

The amount of clearance on a stairway is usually governed only by the desired headroom. It is generally assumed that a stairway with proper headroom automatically has enough clearance. Clearance governs the height of an object which may be moved on the stairs and should be given special consideration when the stair is to be used to move large objects.

Stairwell Length

The total rise, unit rise, and unit run of a stairway are normally established before the length of the stairwell is determined. Next, the desired headroom is established and added to the upper construction. The sum of headroom and upper construction is divided by the unit rise to determine the number of risers included in the length of the stairwell. The result is usually rounded off to the next one-tenth of a riser. However, when additional headroom is desirable or there is no shortage of space for the stairwell, the result may be rounded off to the next full riser. See Figure 3-1.

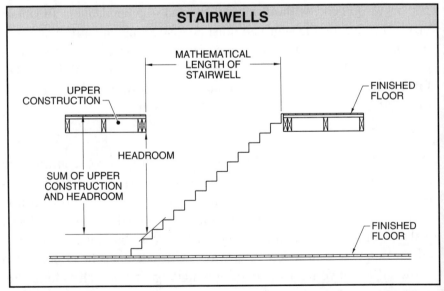

Figure 3-1. The sum of headroom and upper construction is divided by the unit rise to determine the number of risers included in the length of the stairwell.

The number of risers included in the sum of the headroom and upper construction is determined because of the direct relationship between the number of risers included in that distance and the number of treads which must be left in the open, that is, be included in the length of the stairwell. For each riser or fraction of a riser included in the sum of the upper construction and headroom, there is a corresponding tread or fraction of a tread included in the length of the stairwell. Therefore, the number of risers included in the sum of headroom and upper construction is equal to the number of treads left in the open in the length of the stairwell. The minimum length of the stairwell is found by multiplying the number of treads in the open by the unit run. See Figure 3-2.

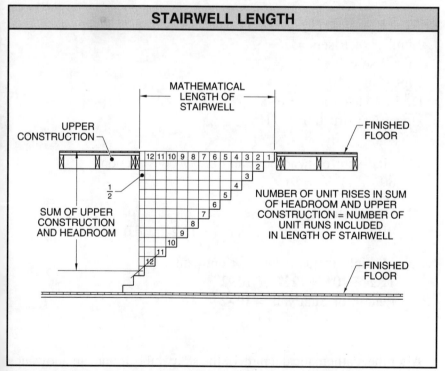

Figure 3-2. The minimum length of the stairwell is found by multiplying the number of treads in the open by the unit run.

40 STAIR LAYOUT

The stairwell length is found by applying the following procedure:

1. Find the sum of the headroom and upper construction.

2. Divide the sum by the unit rise to find the number of treads in the open. *Note:* The number of risers included in the sum of headroom and upper construction equals the number of treads left in the open.

3. Multiply number of treads in the open by the unit run to find stairwell length.

For example, what is the stairwell length for a stairway with the following characteristics?

Total Rise = $112\frac{1}{2}''$
Number of Risers = 15
Unit Rise = $7\frac{1}{2}''$
Unit Run = $10''$
Desired Headroom = $6'\text{-}8''$
Upper Construction = $11\frac{1}{2}''$

1. Find sum of headroom and upper construction.
   $80'' + 11\frac{1}{2}'' = 91\frac{1}{2}''$

2. Divide sum by unit rise.
   $91\frac{1}{2}'' \div 7\frac{1}{2}'' = 12.2$ Risers (= 12.2 treads)

3. Multiply treads in open × unit run.
   $12.2 \times 10'' = 122''$ or $10'\text{-}2''$
   Length of Stairwell = **$10'\text{-}2''$**

After the mathematical length of the stairwell is found, an allowance should be made for the thickness of the riser and finish wall coverings such as plaster or drywall. See Figure 3-3.

Stairwells

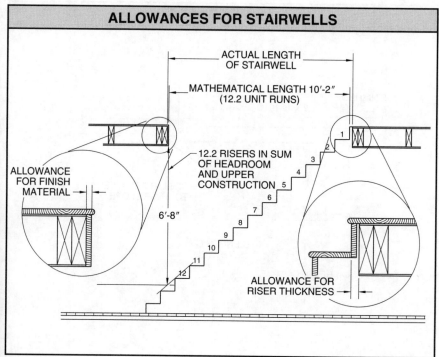

Figure 3-3. An allowance is made for the thickness of the riser and finish wall coverings.

Gaining Usable Space

The floor framework is often slanted to follow the slope of the stair in order to gain usable space over a stairway. This procedure may yield as much as 1'-0" of added floor length over the stairwell and still maintain proper headroom. In this method, 2 × 4s 16" OC (on center) are placed 90° to the joist and 2 bys are placed parallel to the stair slope.

An alternate method used to gain floor space over a stairway in residential work is to place a header flatways over the stairway. This method requires less material and labor than the preceding method. See Figure 3-4. Usable space over a stairway is gained by sloping the floor framework above.

42 STAIR LAYOUT

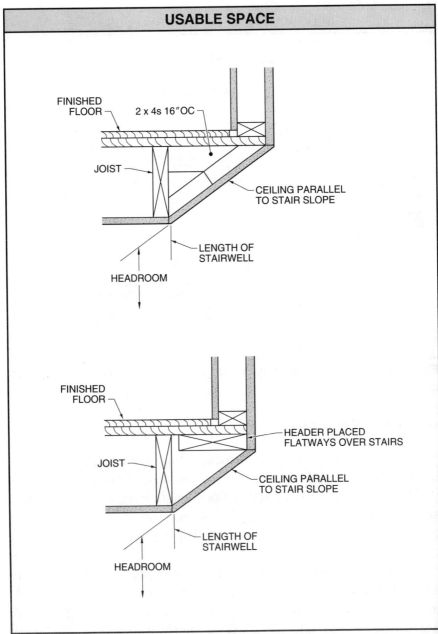

Figure 3-4. Usable space over a stairway is gained by sloping the floor framework above.

Stair Design Steps

In the total design of a stairway, a number of steps must be followed. These steps are:
1. Determine the total rise.
2. Determine the number of risers.
3. Determine the unit rise.
4. Determine the unit run.

Use one of the following formulas:

Formula 1. *Unit Rise plus Unit Run = 17" to 18"*
Formula 2. *Two Unit Rises plus Unit Run = 24" to 25"*
Formula 3. *Unit Rise times Unit Run = 72 to 75*

5. Determine the desired headroom.
6. Determine the thickness of the upper floor construction.
7. Determine sum of the headroom and the upper floor construction.
8. Determine the number of risers and treads to be in the open.
9. Multiply the unit run by the number of treads in the open to find the length of the stairwell.
10. Make an allowance for material thickness in the length of the stairwell.

44 STAIR LAYOUT

TRADE TEST 3
STAIRWELLS

Date _____ Name _____

_____ 1. A headroom height of _____ is acceptable, under most building codes for basement stairways in residential work.

    A. 6'-0"   C. 6'-8"
    B. 6'-4"   D. 7'-0"

_____ 2. Main stairways in residential work should have a minimum headroom of _____.

    A. 6'-0"   C. 6'-8"
    B. 6'-4"   D. 7'-0"

_____ 3. The minimum headroom for main stairways in residential work should be increased to _____ or more whenever possible.

    A. 6'-0"   C. 6'-8"
    B. 6'-4"   D. 7'-0"

T   F   4. The total rise, unit rise, and unit run of a stairway are normally established before the length of the stairwell is determined.

T   F   5. The unit rise plus the unit run of a stairway should equal 17" to 18".

_____ 6. Formula 3 states that *Unit Rise times Unit Run equals* _____.

    A. 68" to 72"   C. either A or B
    B. 72 to 75   D. neither A nor B

46 STAIR LAYOUT

_____ 7. The sum of headroom and _____ is divided by the unit rise to determine the number of risers included in the length of the stairwell.

    A. total run   C. upper construction
    B. lower construction   D. neither A, B, nor C

_____ 8. The minimum length of the stairwell is found by multiplying the number of treads in the open by the _____.

    A. total run   C. either A or B
    B. unit run   D. neither A nor B

_____ 9. Formula 2 states that *Two Unit Rises plus Unit Run equals* _____.

    A. 22" to 24"   C. 22" to 26"
    B. 22" to 25"   D. 22" to 27"

T   F   10. The amount of clearance on a stairway is usually governed only by the desired headroom.

_____ 11. The length of the stairwell at A is _____.

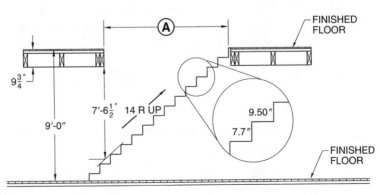

_____ **12.** The length of the stairwell at B is _____.

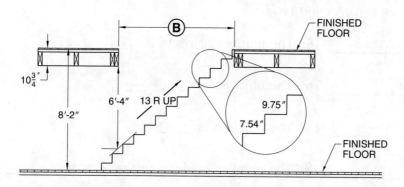

_____ **13.** The length of the stairwell at C is _____.

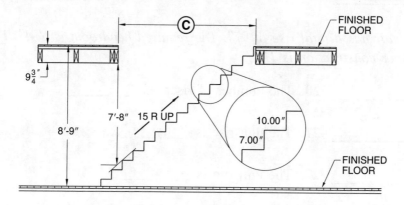

_____ **14.** The length of the stairwell at D is _____.

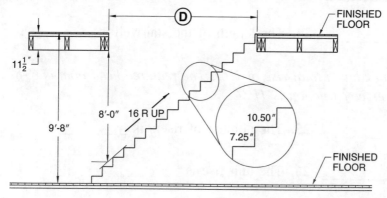

48 STAIR LAYOUT

A stair has a total rise of 110". The required headroom is 6'-8". The upper construction is 11¼".

_____ **15.** The number of risers is _____.

_____ **16.** The unit rise is _____".

_____ **17.** The unit run is _____".

_____ **18.** The number of risers in the open is _____.

_____ **19.** The length of the stairwell is _____.

A stair has a total rise of 97". The required headroom is 6'-4". The upper construction is 11¼".

_____ **20.** The number of risers is _____.

_____ **21.** The unit rise is _____".

_____ **22.** The unit run is _____".

_____ **23.** The number of risers in the open is _____.

_____ **24.** The length of the stairwell is _____.

A stair has a total rise of 116". The required headroom is 6'-8". The upper construction is 11¼".

_____ **25.** The number of risers is _____.

_____ **26.** The unit rise is _____".

_____ 27. The unit run is _____".

_____ 28. The number of risers in the open is _____.

_____ 29. The length of the stairwell is _____.

A stair has a total rise of 127". The required headroom is 7'-0". The upper construction is 11¼".

_____ 30. The number of risers is _____.

_____ 31. The unit rise is _____".

_____ 32. The unit run is _____".

_____ 33. The number of risers in the open is _____.

_____ 34. The length of the stairwell is _____.

A stair has a total rise of 136". The required headroom is 7'-0". The upper construction is 11¼".

_____ 35. The number of risers is _____.

_____ 36. The unit rise is _____".

_____ 37. The unit run is _____".

_____ 38. The number of risers in the open is _____.

_____ 39. The length of the stairwell is _____.

50 STAIR LAYOUT

4 HEADROOM

The total rise, upper construction, length of stairwell, and desired headroom must be known to design stairs with sufficient headroom to fit a stairwell. There are 10 steps to follow when designing stairwells to maintain headroom.

DESIGNING STAIRS TO FIT STAIRWELL

The procedure of designing stairs to fit into an existing stairwell is somewhat different from previously discussed stair layouts. To design this type of stair, the following information must first be known:

Total Rise
Upper Construction
Length of Stairwell
Desired Headroom

Total rise is determined by actually measuring the vertical distance from finished floor to finished floor on the job site. The upper construction and the length of the stairwell are also determined by taking measurements on the job site. When taking the stairwell measurement, allow 1″ for the thickness of the top riser board and shimming. See Figure 4-1. It is good practice to make a scale drawing on grid paper, or with a personal computer and graphics software, to show the existing job conditions and dimensions.

52 STAIR LAYOUT

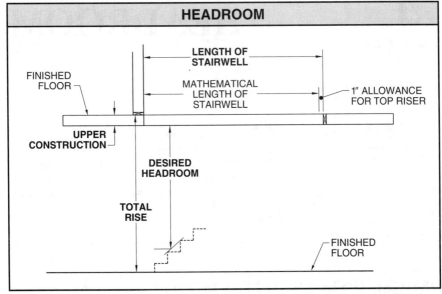

Figure 4-1. The total rise, upper construction, length of stairwell, and desired headroom must be known to design a stairwell with sufficient headroom.

The desired headroom is determined by checking plans and specifications, or if none exist by complying with local building codes and commonly accepted trade practice. Always allow sufficient headroom based upon the type of building in which the stairway is located.

The following procedure is used to design stairs to fit an existing stairwell:

1. Determine total rise. This measurement can be taken on the job site.
2. Determine thickness of upper construction. This measurement can be taken on the job site.
3. Determine length of stairwell. This measurement can be taken on the job site.
4. Determine desired headroom. This should be a minimum of 6'-4" for basements in dwellings and a minimum of 6'-8" for other stairs in a dwelling.

5. Determine number of risers. Divide the finished floor to finished floor dimension by 7″ and drop the remainder, or divide the finished floor to finished floor dimension by 8″ and round off to the next full riser.
6. Determine unit rise. Divide total rise by number of risers.
7. Determine sum of headroom and thickness of upper construction.
8. Determine number of risers and treads in the open. Divide the sum of the headroom and thickness of upper construction by the unit rise.
9. Determine unit run. Divide the length of the stairwell by the number of treads in the open.
10. Check solution with stair formulas.

For example, given the following, determine each solution.

Total Rise = 105″
Upper Construction = 9″
Length of Stairwell = 107″ (1″ allowance already made for riser thickness)
Desired Headroom = 80″

1. Determine total rise.
   105″
2. Determine thickness of upper construction.
   9″
3. Determine length of stairwell.
   107″
4. Determine desired headroom.
   80″
5. Determine number of risers.
   105″ ÷ 8″ = 13+ or **14**
6. Determine unit rise.
   105″ ÷ 14 = **7.5″**
7. Determine sum of headroom and thickness of upper construction.
   80″ + 9″ = **89″**
8. Determine number of risers and treads in the open.
   89″ ÷ 7.5″ = 11.86 or **11.9**

54 STAIR LAYOUT

9. Determine unit run.
   107″ ÷ 11.9 = 8.99 or **9″**
10. Check solution with stair formulas.
    Formula 2: *Two Unit Rises + Unit Run = 24″ to 25″*
    2 × 7.5″ + 9″ = 15″ + 9″ = **24″**

This solution has a satisfactory unit rise (7.5″) and unit run (9″), and it gives a stair with the desired headroom (80″).

When the unit run obtained does not meet the requirements of the stair ratio formulas, the stair solution should be adjusted to comply with one of the formulas. The first change to make is to reduce the number of risers and thereby increase the unit rise. If the new unit rise is acceptable, it may be used with the previously determined unit run or a new unit run may be calculated.

For example, given the following, determine each solution.

Total Rise = 8′-9¾″ = 105¾″
Upper Construction = 9¾″
Length of Stairwell = 10′-2¼″ = 122¼″ (1″ allowance already made for riser thickness)
Desired Headroom = 6′-8″ = 80″

1. Determine total rise.
   105¾″
2. Determine thickness of upper construction.
   9¾″
3. Determine length of stairwell.
   122¼″
4. Determine desired headroom.
   80″
5. Determine number of risers.
   105.75″ ÷ 7 = 15+ or **15**
6. Determine unit rise.
   105.75″ ÷ 15 = **7.05″**
7. Determine sum of headroom and thickness of upper construction.
   80″ + 9¾″ = **89¾″**

8. Determine number of risers and treads in the open.
   89.75″ ÷ 7.05″ = 12.73 or **12.8**
9. Determine unit run.
   122.25″ ÷ 12.8 = **9.55″**
10. Check solution with stair formulas.
    Formula 2: *Two Unit Rises + Unit Run = 24″ to 25″*
    2 × 7.05″ + 9.55″ = 14.10″ + 9.55″ = **23.65″**

Solution 1 does not check with the stair ratio formula, and since the unit run cannot be increased, the unit rise must be increased if possible. To increase the unit rise, the number of risers is decreased by one to 14 risers and the headroom is increased by 6.89″.

To find the increased headroom, multiply the number of riser and treads in the open (12.8) by the new unit rise (7.55″) to find the new sum of headroom plus upper construction (12.8 × 7.55″ = 96.64″). The new sum of headroom plus upper construction (96.64″) minus the old sum of headroom plus upper construction (89.75″) = 6.89″ (96.64″ − 89.75″ = 6.89″).

Steps 5 – 10 are worked again (Solution 2), using the new number of risers and headroom.

5. Determine number of risers.
   105.75″ ÷ 7 = 15+ or 15 . . . changed to **14**
6. Determine unit rise.
   105.75″ ÷ 14 = **7.55″**
7. Determine sum of headroom and thickness of upper construction.
   86.89″ + 9¾″ = **96.64″**
8. Determine number of risers and treads in the open.
   96.64″ ÷ 7.55″ = **12.8**
9. Determine unit run.
   122.25″ ÷ 12.8 = **9.55″**
10. Check solution with stair formulas.
    Formula 2: *Two Unit Rises + Unit Run = 24″ to 25″*
    2 × 7.55″ + 9.55″ = 15.1″ + 9.55″ = **24.65″**

Using the new unit rise (7.55″) with the previously determined unit run (9.55″) in the stair formula results in a satisfactory stair ratio. Solution 2 also gives added headroom of 6.89″. See Figure 4-2.

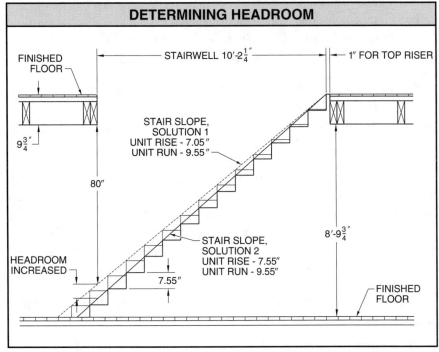

Figure 4-2. The stair solution for determining headroom may be adjusted to comply with one of the formulas.

A third solution to this problem, in which the desired headroom is maintained, is possible. In this solution, a new unit run is determined to go along with the newly established unit rise and Steps 8 – 10 are worked again.

8. Determine number of risers and treads in the open.
   89.75″ ÷ 7.55″ = 11.89 or **11.9**
9. Determine unit run.
   122.25″ ÷ 11.9 = **10.27″**

10. Check solution with stair formulas.
   Formula 2: *Two Unit Rises + Unit Run = 24" to 25"*
   2 × 7.55" + 10.27" = 15.1" + 10.27" = **25.37"**

Step 10 shows that the newly determined unit rise (7.55") and run (9.55") do not fit the stair ratio in Formula 2. However, it does fit the stair ratio in Formula 1: *Unit Rise + Unit Run = 17" to 18"* as shown when Step 10 is worked again using Formula 1.

10. Check solution with stair formulas.
    Formula 1: *Unit Rise + Unit Run = 17" to 18"*.
    7.55" + 10.27" = **17.82"**

Since the last solution results in a stairway with a flatter slope, it may be the most desirable.

FINDING ACTUAL HEADROOM

Occasionally it is necessary to determine the actual headroom of a given stair design without the aid of a scale drawing. However, it is possible to determine the actual headroom of a stairway if the following facts are known: total rise, unit rise, unit run, upper construction, and length of stairwell.

The following procedure is used to determine the actual headroom:
1. Determine number of treads and risers in the open. Divide the length of the stairwell by the unit run.
2. Determine sum of headroom and upper construction. Multiply the number of risers in the open by the unit rise.
3. Determine actual headroom. Subtract the thickness of the upper construction from the sum of headroom and upper construction.

For example, given the following, determine the actual headroom.
Total Rise = 9'-4½"
Unit Rise = 7½"
Unit Run = 10"
Upper Construction = 11½"
Length of Stairwell = 10'-5"

58 STAIR LAYOUT

1. Determine number of treads and risers in the open.
   $125'' \div 10'' = \mathbf{12.5}$
2. Determine sum of headroom and upper construction.
   $12.5 \times 7.5'' = \mathbf{93.75''}$
3. Determine actual headroom.
   $93.75'' - 11.5'' = \mathbf{82.25''}$

In this type of problem, the total rise of the stair should be known mainly for checking the results. If the actual headroom found by the preceding methods exceeds the total rise of the stair minus the thickness of the upper construction, the maximum headroom at the bottom of the stair is governed by the ceiling height at the bottom of the stair.

This point is illustrated by using a total rise of 90″ in the preceding problem. Given the following, determine the actual headroom.

Total Rise = 90″
Unit Rise = 7½″
Unit Run = 10″
Upper Construction = 11½″
Length of Stairwell = 10′-5″

1. Determine number of treads and risers in the open.
   $125'' \div 10'' = \mathbf{12.5}$
2. Determine sum of headroom and upper construction.
   $12 \times 7.5'' = \mathbf{90.00''}$
3. Determine actual headroom.
   $90.00'' - 11.5'' = \mathbf{78.5''}$

The headroom determined was 82¼″. However, with a 90″ total rise which includes 11½″ of upper construction, the lower ceiling height would only be 78½″ which would be the headroom. It should also be noted that there could have been 12½ risers in the open, but with a 90″ total rise, the stair would have only 12 risers. Therefore, if the number of risers in the open is equal to or greater than the number of risers in the stairs, it immediately becomes apparent that the headroom is governed by the ceiling height at the bottom of the stair.

Headroom Design Steps

The steps used in designing a stair to meet existing conditions are simple and are generally used in the following sequence.
1. Determine the total rise.
2. Determine the thickness of upper construction.
3. Determine the length of stairwell.
4. Determine the desired headroom.
5. Determine the number of risers.
6. Determine the unit rise.
7. Determine sum of the headroom and thickness of upper construction.
8. Determine the number of risers and treads in the open.
9. Determine the unit run.
10. Check solution with stair formulas.

TRADE TEST 4
HEADROOM

Date _____ Name _____

T   F   **1.** Total rise may be determined by measuring the vertical distance from finished floor to finished floor on the job site.

_____   **2.** The desired headroom is determined by _____.

     A. checking plans and specifications
     B. complying with local building codes
     C. following trade practices
     D. A, B, and C

_____   **3.** The minimum headroom for a basement stair in a dwelling is _____.

_____   **4.** The minimum headroom for a stair (other than basement) in a dwelling is _____.

     A. 6′-4″   C. 7′-0″
     B. 6′-8″   D. neither A, B, nor C

_____   **5.** To find unit rise, _____ total rise by the number of risers.

     A. multiply   C. multiply ½ the
     B. divide   D. neither A, B, nor C

T   F   **6.** The length of the stairwell may be determined by taking a measurement on the job site.

T   F   **7.** The unit run may be determined by taking a measurement on the job site.

62 STAIR LAYOUT

_____ 8. To determine the number of risers, divide the finished floor to finished floor dimension by _____ " and drop the remainder.

_____ 9. To determine the number of risers, divide the finished floor to finished floor dimension by _____ " and round off to the next full riser.

_____ 10. To determine the number of risers and treads in the open, divide the sum of the headroom and thickness of upper construction by the _____.

Use Stairway A for Problems 11 through 15.

_____ 11. The number of risers is _____.

_____ 12. The unit rise is _____ ".

_____ 13. The sum of the headroom and upper construction is _____ ".

_____ 14. The number of risers in the open is _____.

_____ 15. The unit run is _____ ".

FINISHED FLOOR

LENGTH OF STAIRWELL 9'-5"

UPPER CONSTRUCTION $11\frac{3}{4}$"

DESIRED HEADROOM 6'-8"

TOTAL RISE $9'-0\frac{3}{4}$"

FINISHED FLOOR

STAIRWAY A

Use Stairway B for Problems 16 through 20.

_____ **16.** The number of risers is _____.

_____ **17.** The unit rise is _____ ″.

_____ **18.** The sum of the headroom and upper construction is _____ ″.

_____ **19.** The number of risers in the open is _____.

_____ **20.** The unit run is _____ ″.

FINISHED FLOOR

LENGTH OF STAIRWELL 9′-8″

UPPER CONSTRUCTION $11\tfrac{3}{4}″$

DESIRED HEADROOM 6′-8″

TOTAL RISE $8′\text{-}7\tfrac{1}{4}″$

FINISHED FLOOR

STAIRWAY B

Use Stairway C for Problems 21 through 25.

_____ **21.** The number of risers is _____.

_____ **22.** The unit rise is _____ ″.

_____ **23.** The sum of the headroom and upper construction is _____ ″.

_____ **24.** The number of risers in the open is _____.

_____ **25.** The unit run is _____ ″.

64 STAIR LAYOUT

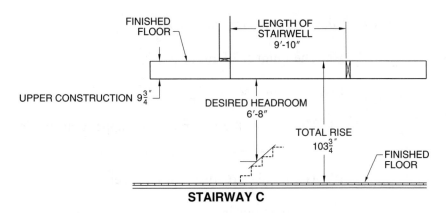

STAIRWAY C

Use Stairway D for Problems 26 through 30.

_____ 26. The number of risers is _____.

_____ 27. The unit rise is _____".

_____ 28. The sum of the headroom and upper construction is _____".

_____ 29. The number of risers in the open is _____.

_____ 30. The unit run is _____".

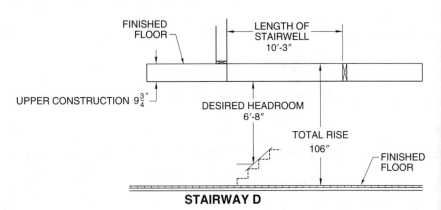

STAIRWAY D

Use Stairway E for Problems 31 through 35.

_____   **31.** The number of risers is _____.

_____   **32.** The unit rise is _____".

_____   **33.** The sum of the headroom and upper construction is _____".

_____   **34.** The number of risers in the open is _____.

_____   **35.** The unit run is _____".

FINISHED FLOOR

LENGTH OF STAIRWELL 9'-8"

UPPER CONSTRUCTION $9\frac{3}{4}"$

DESIRED HEADROOM 6'-8"

TOTAL RISE 110"

FINISHED FLOOR

STAIRWAY E

66 STAIR LAYOUT

5 LANDINGS

The design and layout of stairs requiring landings follows the same general procedure employed in the design and layout of plain straight flight stairs. The additional element introduced is the necessary allowance for the size of the landing.

LANDINGS

A landing is a platform separating two flights of stairs. Landings in long flights of stairs serve as resting places for persons climbing the stairs and also serve to break the fall of anyone slipping and falling on the stairway.

Landings may be incorporated into any type of stairway. Landings used in any stair should be at least as wide as the stair itself. A landing installed to break a long flight of stairs should measure at least 36″ in the direction of the line of travel. Landings are installed in stairways with over 22 risers so that no single flight of stairs contain more than 18 risers or less than 3 risers. See Figure 5-1.

Once the need for a landing is determined, the location of the landing can be established. In residential work, the landing is usually installed halfway between the two floors, or if more than one landing is used, they are usually equally spaced.

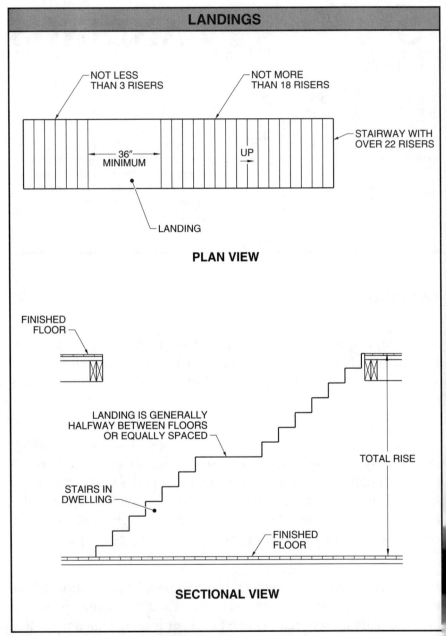

Figure 5-1. An unbroken flight of stairs should not contain more than 18 risers nor less than three risers.

Landings in Straight Stairs

The initial steps of designing landings in straight stairs are the same as that of designing an unbroken, straight flight of stairs. The procedure is:

1. Determine number of risers. Divide total rise by 7" and drop the remainder or divide total rise by 8" and round off to the next full riser.
2. Determine unit rise. Divide total rise by the number of risers.
3. Determine height of landing.

First, the total rise of the stair is determined. Next, the number of risers and the unit rise are calculated. If the number of risers is even (14, 16, 18, etc.), a single landing can be located halfway between the floors with a uniform unit rise from floor to floor. If the number of risers is odd (15, 17, 19, etc.), the landing is placed nearer one floor in order to maintain a uniform unit rise for the entire stairway. See Figure 5-2.

For example, locate a landing in a straight stairway having a total rise of 12'-1".

1. Determine number of risers.
   145" ÷ 7" = 20+ or **20**
2. Determine unit rise.
   145" ÷ 20 = **7.25"**
3. Determine height of landing. Since there is an even number of risers, divide the number of risers by 2 and multiply times the unit rise.
   20 ÷ 2 = 10
   10 × 7.25" = 72.5"
   Height of Landing = **72.5"**

70 STAIR LAYOUT

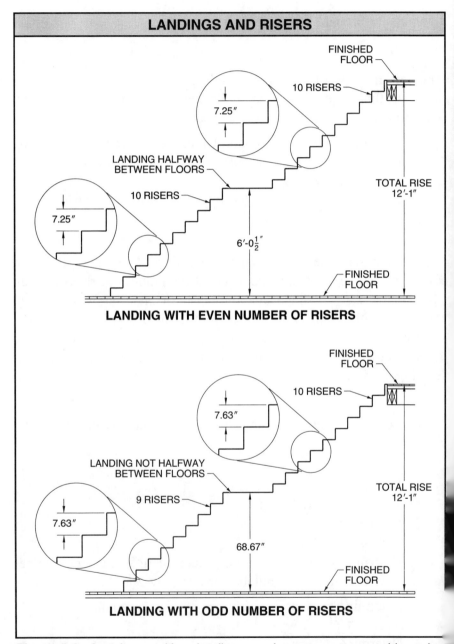

Figure 5-2. A stairway with a landing may have an even or odd number of risers.

An alternate solution to this same problem is:
1. Determine number of risers.
   145″ ÷ 8″ = 18+ or **19**
2. Determine unit rise.
   145″ ÷ 19 = **7.63″**
3. Determine height of landing. Since there is an odd number of risers, locate the landing 9 risers above the lower floor and 10 risers below the upper floor. *Note:* This could also be reversed.
   9 × 7.63″ = 68.67″
   10 × 7.63″ = 76.3″
   Height of Landing = **68.67″**

If it is required that the landing be placed halfway between floors where there is an odd number of risers, the landing could be located first and the distance from the landing to each floor treated as a separate total rise. See Figure 5-3.

For example, locate a landing halfway between floors in a straight stair with a total rise of 192.5″.
1. Determine number of risers. *Note:* Find the number of equal risers for the entire stair.
   192.5″ ÷ 7″ = 27+ or **27**

Since this results in an odd number of risers, the landing cannot be placed halfway between floors with the unit rise based on this solution. This problem must be treated as two separate stairways.

2. Determine total rise of each stair.
   192.5″ ÷ 2 = **96.25″**

Note: This dimension is the vertical distance from the floors to the landing.

3. Determine number of risers from floors to landing.
   96.25″ ÷ 7″ = 13+ or **13**
4. Determine unit rise.
   96.25″ ÷ 13 = **7.40″**

72 STAIR LAYOUT

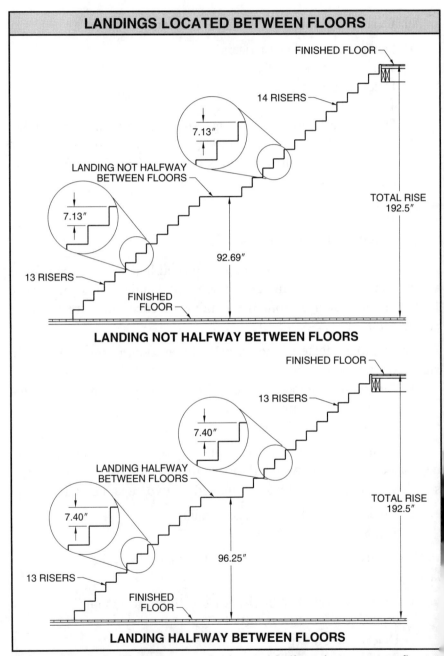

Figure 5-3. A single landing can be located halfway between two floors.

After the location of the landing or landings is known and the unit rise determined, they may be marked off on a story pole by the stair builder. A *story pole* is a strip of lumber, usually a 1 × 2 or 2 × 2, on which the total rise of the stair is indicated.

The unit run for straight stairs with a landing is calculated by using any one of the stair ratio formulas. Required headroom is determined by checking the plans and specifications or the building code or following commonly accepted trade practice. The thickness of the upper construction is determined by checking a sectional drawing of the floor construction.

To find the length of the stairwell, the sum of the headroom and upper construction thickness is first divided by the unit rise. The result is the number of risers and treads in the open. The actual length of the stairwell is now found by multiplying the number of treads in the open by the unit run and making the necessary allowance for the length of the landing if it falls in the open. See Figure 5-4.

For example, design a stairway with a total rise of 16'-0½" and place a 36" landing halfway between floors. The required headroom is 7'-6", and the thickness of the upper floor construction is 12¼".

1. Determine total rise of each stair.
   16'- 0½" ÷ 2 = **8'-0¼"**
2. Determine number of risers between floor and landing.
   96¼" ÷ 7" = 13+ or **13**
3. Determine unit rise.
   96¼" ÷ 13 = **7.40"**
4. Determine unit run. *Note:* Formula 1 chosen for this example.
   17½" − 7.40" = **10.10"**
5. Determine sum of headroom and upper construction.
   90" + 12¼" = **102¼"**

74 STAIR LAYOUT

6. Determine number of risers and treads to be left in the open.
$102\frac{1}{4}" \div 7.40" = 13.8+$ or **14**
7. Determine length of stairwell. Since 14 treads must be left in the open and there are only 12 treads above the landing, the length of the landing makes up the width of one tread.
$13 \times 10.1" = 131.30"$
Length of Landing = 36.00"
Length of Stairwell = $131.30" + 36.00" = 167.3"$ or **$13'\text{-}11\frac{5}{16}"$**

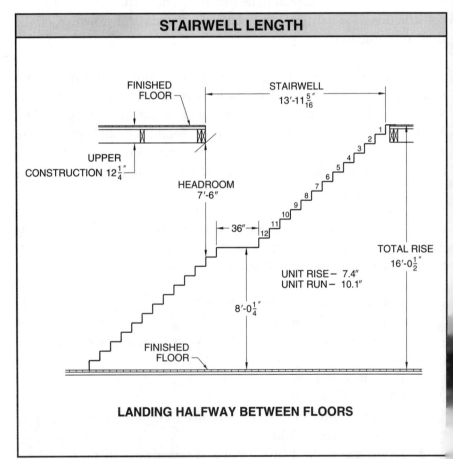

Figure 5-4. The actual length of the stairwell is found by multiplying the number of treads in the open by the unit run.

Quarter-turn Stairs

A quarter-turn stair has a landing where the stairway changes directions 90°. This change in direction is usually made necessary by the lack of sufficient space for a straight stairway. No general rule can be followed as to the placement of the landing in relation to the total number of risers in the stair. In the initial design of a quarter-turn stair, the landing is located by a trial and error method.

The procedure applied in the design of this type of stair follows a pattern similar to that used in previous stair problems. First, the total rise is determined. Next, the number of risers and the unit rise are determined. The unit run is established with the aid of one of the stair ratio formulas.

The landing is located in the plan view by determining how many unit runs can be placed in each direction based on the space available. After the landing is located in the plan view, the height is established similar to the height for straight stairs with landings.

The dimensions of the stairwell are determined in conjunction with the plan view and the required headroom along with the upper construction thickness. The number of risers and treads in the open is determined by dividing the sum of the headroom and upper construction thickness by the unit rise. With the aid of the plan view, the number of risers and treads below the landing which must remain in the open is determined.

The sum of the unit runs below the landing plus the length of the landing gives the length of the stairwell in one direction. The sum of the unit runs above the landing plus the width of the landing gives the length of the stairwell in the other direction. See Figure 5-5.

76 STAIR LAYOUT

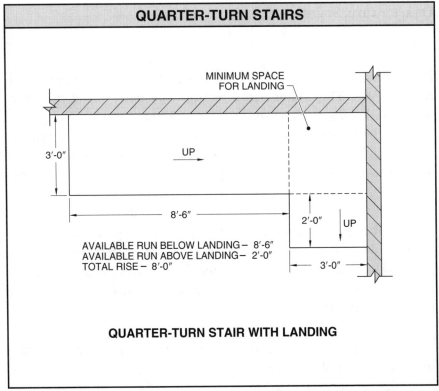

Figure 5-5. The landing for a quarter-turn stair is located in the plan view by determining how many unit runs can be placed in each direction.

For example, design a 3'-0" wide quarter-turn stair with a total rise of 8'-10". The available run below the landing is 8'-6". The available run above the landing is 2'-0".
1. Determine total rise. *See Plan.*
   Total Rise = **8'-10"**
2. Determine number of risers.
   106" ÷ 8" = 13+ or **14**
3. Determine unit rise.
   106" ÷ 14 = **7.57"**

4. Determine unit run. *Note:* Formula 1 chosen for this example. The unit run in this example could be increased or decreased by any amount up to ½″ and still fit the stair ratio.
   17.5″ − 7.57″ = **9.93″**
5. Determine number of treads below the landing. Available run below landing is 8′-6″ or 102″.
   102″ ÷ 9.93″ = 10+ or **10**
6. Determine height of landing above first floor. *Note:* If there are 10 treads below the landing, there will be 11 risers below the landing.
   11 × 7.57″ = **83.27″**
7. Determine number of risers above the landing.
   14 − 11 = **3**
8. Determine run above the landing.
   2 × 9.93″ = **19.86″**
9. Lay out plan view of stairway.

The plan view should be arranged so that the nosing of the lower tread in the upper stair does not project past the face of the stringer of the lower stair. See Figure 5-6. After the plan view for the stair is completed, the length of the stairwell is determined as follows:

10. Determine sum of headroom and upper construction.
    6′-8″ + 10″ = 80″ + 10″ = **90″**
11. Determine number of risers and treads in the open.
    90″ ÷ 7.57″ = 11.8+ or **11.9**
12. Determine number of risers and treads below landing in the open.
    11.9 − 3 = **8.9**
13. Determine length of stairwell in the direction below landing.
    8.9 × 9.93″ = 88.38″
    Length of Landing = 38.00″
    Length of Stairwell = 88.38″ + 38.00″ = 126.38″ or **10′-6⅜″**

78 STAIR LAYOUT

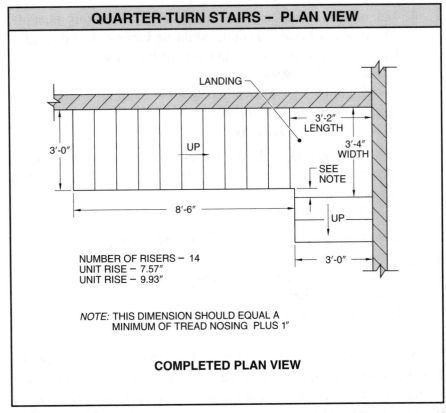

Figure 5-6. The plan view is arranged so that the nosing of the lower tread in the upper stair does not project past the face of the stringer of the lower stair.

14. Determine length of stairwell in the direction above landing.
    2 × 9.93″ = 19.86″
    Width of Landing = 40.00″
    Length of Stairwell = 19.86″ + 40.00″ = 59.86″ or **4′-11⅞″**

A check of the Plan View shows that this solution is possible. The completed layout, in Sectional View, shows the required headroom along with the number of risers and treads in the open. See Figure 5-7.

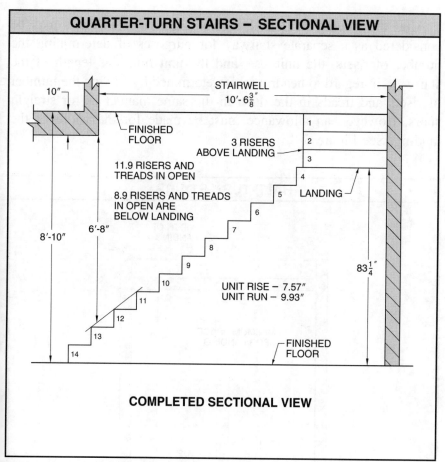

Figure 5-7. The sectional view shows the required headroom along with the number of risers and treads in the open.

Half-turn Stairs

The half-turn stair is made up of two straight flights of stairs with a landing installed where the stairway changes directions 180°. The landing is usually installed halfway between the two floors although it may be placed nearer one floor if job conditions required it.

80 STAIR LAYOUT

After the landing is located, each straight flight of stairs may be considered as a separate stairway for purposes of determining the number of risers, the unit rise, and the unit run. The length of the stairwell, in regard to headroom, is determined by finding the number of risers and treads in the open in the same manner as for straight stairs. However, an allowance must be made for the width of the landing. See Figure 5-8.

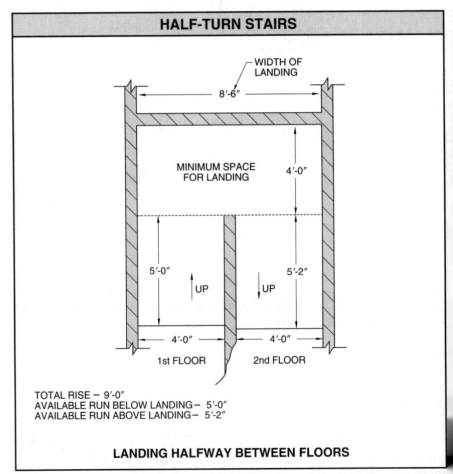

Figure 5-8. The landing for a half-turn stair is usually placed halfway between the two floors.

For example, design a 4'-0" wide half-turn stair with a total rise of 9'-0". The required headroom is 6'-8" with an upper construction of 12". The available run below the landing is 5'-0". The available run above the landing is 5'-2". *Note:* The landing will be located halfway between floors because it has a nearly equal amount of space on each side for the steps. See Figure 5-9.

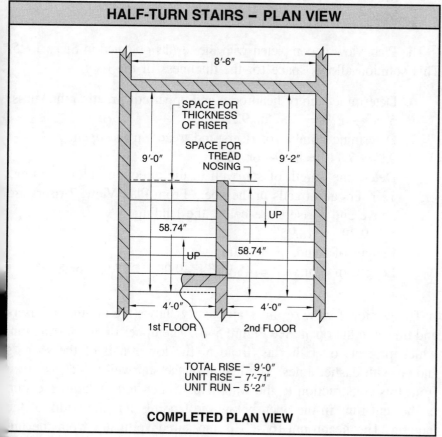

Figure 5-9. The plan view is completed when the height of the landing, number of risers per flight, unit rise, and unit run have been determined.

1. Determine height of landing.
   $9'\text{-}0'' \div 2 = \mathbf{4'\text{-}6''}$
2. Determine number of risers in each flight.
   $54'' \div 7'' = 7+$ or $\mathbf{7}$
3. Determine unit rise.
   $54'' \div 7 = \mathbf{7.71''}$
4. Determine unit run. *Note:* Formula 1 chosen for this example.
   $17.50'' - 7.71'' = \mathbf{9.79''}$
5. Check results against Plan View.
   Maximum space allowed for treads = $60''$
   Space required by treads = $9.79'' \times 6 = \mathbf{58.74''}$

The Plan View is completed with the results obtained in Steps 1 – 5. This solution allows space for the thickness of the riser.

6. Determine sum of headroom and upper construction thickness.
   $6'\text{-}8'' + 12'' = 7'\text{-}8''$ or $\mathbf{92''}$
7. Determine number of risers and treads in the open.
   $92'' \div 7.71'' = 11.9+$ or $\mathbf{12}$
8. Determine length of stairwell in the direction below landing.
   12 risers and treads in the open. From Plan View, 7 risers are above and 5 risers are below the landing
   $5 \times 9.79'' = 48.95''$
   Length of Landing = $49.00''$
   Length of Stairwell = $48.95'' + 49.00'' = 97.95''$ or $\mathbf{8'\text{-}2''}$

The sectional view of this stair clearly shows the number of risers and treads in the open. See Figure 5-10. The upper floor construction, which projects over the last tread of the lower half of the stair, is shown with dashed lines. The length of the stairwell ($8'\text{-}2''$) is given from this construction to the wall. This dimension includes the sum of the unit runs in the open below the landing plus the width of the landing. The headroom ($6'\text{-}8''$) is measured vertically from the underside of the upper floor construction to the slope of the stair.

Landings 83

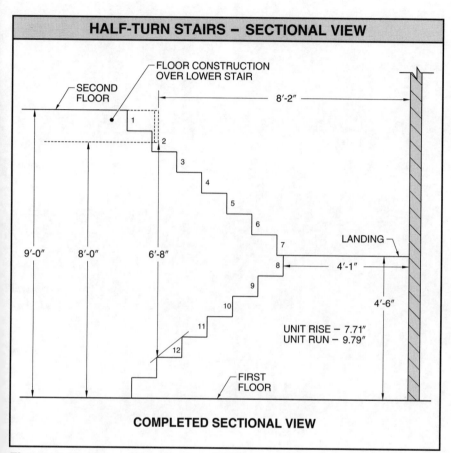

Figure 5-10. The sectional view clearly shows the number of risers and treads in the open.

84 STAIR LAYOUT

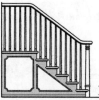

TRADE TEST 5
LANDINGS

Date _____ Name _____

_____ 1. A landing is a(n) _____ separating two flights of stairs.

_____ 2. Landings are installed in stairways with over _____ risers.
   A. 3   C. 22
   B. 18   D. 36

_____ 3. A landing should measure at least _____ " in the direction of travel.
   A. 3   C. 22
   B. 18   D. 36

_____ 4. A single flight of stairs in stairs with a landing should contain no more than _____ risers.
   A. 3   C. 22
   B. 18   D. 36

_____ 5. A single flight of stairs in stairs with a landing should contain no less than _____ risers.
   A. 3   C. 22
   B. 18   D. 36

_____ 6. To determine the number of risers, divide the total rise by _____ " and drop the remainder.

_____ 7. To determine the number of risers, divide the total rise by _____ " and round off to the next full riser.

85

86 STAIR LAYOUT

Use Stairway A (Straight Stairway with 36" Central Landing) for Problems 8 through 12.

_____ 8. The landing is located _____" above the lower floor.

_____ 9. The total number of risers is _____.

_____ 10. The unit rise is _____".

_____ 11. The unit run is _____".

_____ 12. The length of the stairwell is _____.

FINISHED FLOOR
UPPER CONSTRUCTION 16"
DESIRED HEADROOM 7'-0"
TOTAL RISE 16'-2"
FINISHED FLOOR

STAIRWAY A

Use Stairway B (Straight Stairway with 36" Central Landing) for Problems 13 through 17.

_____ 13. The landing is located _____" above the lower floor.

_____ 14. The total number of risers is _____.

_____ 15. The unit rise is _____".

16. The unit run is _____".

17. The length of the stairwell is _____.

```
                    FINISHED
                    FLOOR
                      |
    ━━━━━━━━━━━━━━━━━━━━━━━━━━━━━━━━━━━━━━━
UPPER CONSTRUCTION 16"
              DESIRED HEADROOM
                    7'-0"
                              TOTAL RISE
                               14'-8"
                                           FINISHED
                                           FLOOR
```

STAIRWAY B

Use Stairway C (Quarter-Turn Stairs with 3 Risers Below 2nd Floor) for Problems 18 through 21. Note: The unit rise and unit run equals 17".

18. The total number of risers is _____.

19. The unit rise is _____".

20. The unit run is _____"

21. The height of the landing is _____.

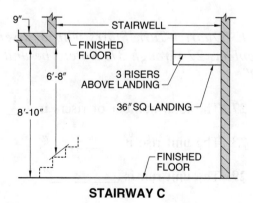

STAIRWAY C

88 STAIR LAYOUT

Use Stairway D (Quarter-Turn Stairs with 3 Risers Below 2nd Floor) for Problems 22 through 26. Note: The unit rise and unit run equals 17".

_____ 22. The total number of risers is _____.

_____ 23. The unit rise is _____".

_____ 24. The unit run is _____"

_____ 25. The height of the landing is _____.

_____ 26. The length of the stairwell is _____.

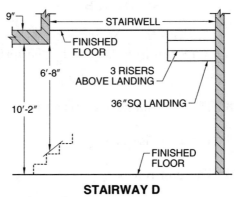

STAIRWAY D

Use Stairway E (Quarter-Turn Stairs with 3 Risers Below 2nd Floor) for Problems 27 through 31. Note: The unit rise and unit run equals 17".

_____ 27. The total number of risers is _____.

_____ 28. The unit rise is _____".

_____ 29. The unit run is _____"

30. The height of the landing is _____.

_____ 31. The length of the stairwell is _____.

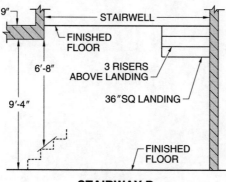

STAIRWAY D

90 STAIR LAYOUT

6 STRINGERS

Stringers (carriages) are the 2 by (most common) boards containing the rise and run of a stair. These include the cut-out carriage, built-up carriage, dadoed stringer, cleated stringer, housed stringer, and cut and mitered stringer. The framing square is used to layout stringers.

STRINGERS

The location and desired finished appearance of stairs govern the species and grade of lumber used. See Appendix. As a general rule, stairs that are exposed to the weather should be made from lumber which can withstand the effects of sun, rain, and snow. Yellow pine, white pine, red fir, and white oak are among the lumber species used for exterior stairs. Many building codes require that exterior stairs be constructed of lumber that has been treated to make it decay resistant.

Interior stairs have been built of most kinds of hardwoods and softwoods commercially available. In selecting material for a stairway, durability and resistance to wear should be given prime consideration, especially when selecting material for the stair treads.

There are several types of stair stringers (carriages) which may be used to support a stairway. These include the cut-out carriage, built-up carriage, dadoed stringer, cleated stringer, housed stringer, and cut and mitered stringer. The type of stringer used depends on the location of the stairway and the desired finished appearance.

Cut-out Carriages

The cut-out carriage is one of the most common types of stringers. See Figure 6-1. Depending on the slope and location of the stairs, this carriage may be made from 2 × 10s, 2 × 12s, 2 × 14s, or even larger stock. The 2 × 10 is the most common size of lumber used for cut-out carriages in residential work.

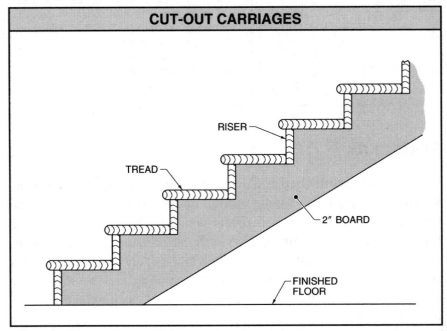

Figure 6-1. The cut-out carriage is one of the most common types of stringers.

Built-up Carriages

A variation of the cut-out carriage is the built-up carriage, which consists of a 2 × 4 or 2 × 6 member with blocks fastened to it. See Figure 6-2. These blocks are often the material which has been cut from a board used to make a cut-out stringer. Using the blocks cut from a cut-out carriage to make a built-up carriage with a 2 × 4 or 2 × 6 is an economical way of obtaining a second carriage. The connection may be made by nailing the blocks to the supporting stock or by nailing cleats to the side of the blocks and the supporting member.

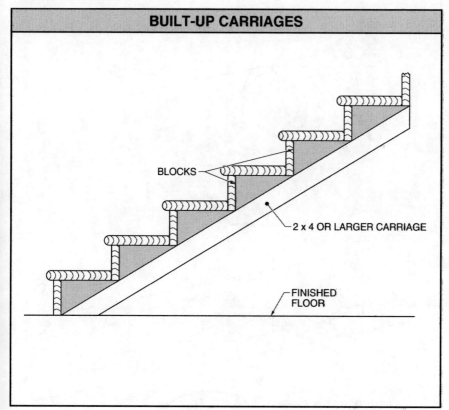

Figure 6-2. The built-up carriage is a variation of the cut-out carriage.

Layout of Stair Carriages. Stairs are often laid out with the aid of stair gauges clamped to the square. These gauges are clamped on the square so that the unit rise and unit run fall along the edge of the carriage stock or along a layout line drawn on the stock. See Figure 6-3.

94 STAIR LAYOUT

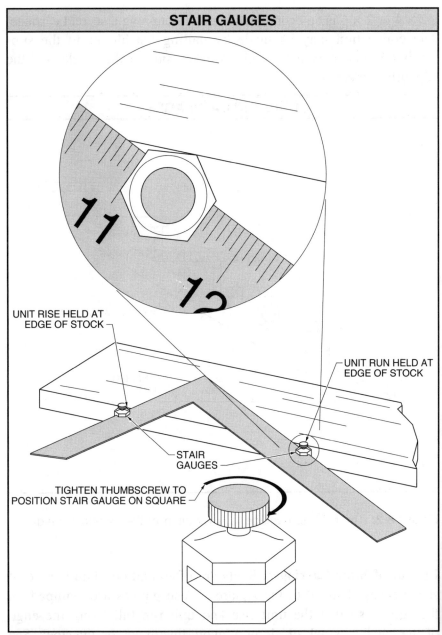

Figure 6-3. Stair gauges are clamped on the square at the unit rise and unit run dimensions.

The unit rise and unit run must be determined before stair carriages are laid out. Stair gauges are clamped on the square with the rise on the tongue of the square and the run on the body of the square. The board selected for the carriage is laid on a pair of saw horses or on a workbench of convenient height with the top edge of the material toward the layout person. When placed in this manner, the upper end of the stringer is on the layout person's left. Layout proceeds from left to right. See Figure 6-4.

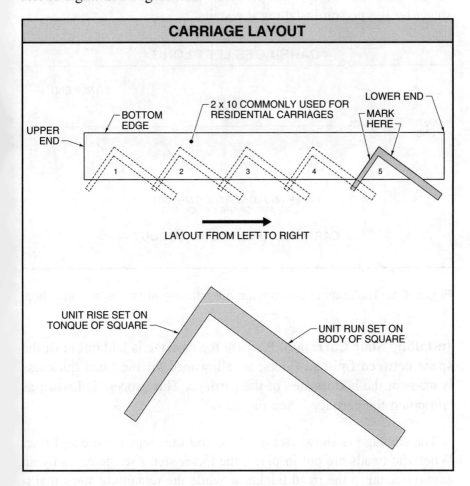

Figure 6-4. Layout on a stair carriage proceeds from left to right.

96 STAIR LAYOUT

A sharp pencil is used to mark the steps as a small error on each step can amount to a large error in the completed carriage. It is good practice to number the risers as each step is marked to avoid the possibility of miscounting and skipping a step. Number each riser as it is marked to avoid this costly mistake.

As a general rule, when carriages are made in the shop, only the shape of the steps is cut out. The upper and lower ends of the carriage are left long. See Figure 6-5. The ends are cut to the proper size and shape by the person installing the stairs.

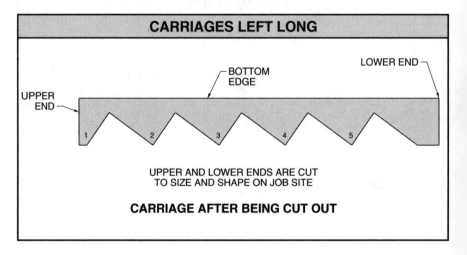

Figure 6-5. The ends of the carriage are left long when made in the shop.

Installing Stair Carriages. Because the carriage is laid out to fit the space between finished floors, an allowance for the tread thickness is made at the bottom step of the carriage. This process is known as "dropping the carriage." See Figure 6-6.

The carriage is shown set in place and the steps have equal rise. When the treads are put in place, the lower step rise increases by an amount equal to the tread thickness while the remaining steps maintain the same height.

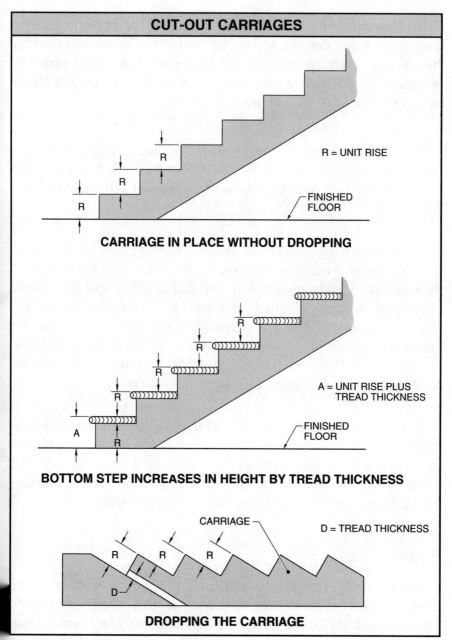

Figure 6-6. Dropping the carriage assures that all risers are the same height.

To compensate for the additional height of the bottom step, the lower end of the carriage is laid out and cut. The amount to drop the carriage is always equal to the thickness of the tread material. Dropping the carriage in this manner results in a stairway with an equal unit rise from the bottom to the top.

Fitting carriages between two walls allows the carriages to be supported by fastening them to the walls, but when there are no side walls, the carriages must be fastened securely to the floor frame. See Figure 6-7.

A 2 × 12 carriage is cut out to allow the top riser to be placed on blocking or wedges against the header joist. The lower part of the carriage extends upward and is nailed to the sides of the joists or to solid bridging installed for the purpose of fastening the carriages. If additional strength is required, a 2 × 2 or 2 × 4 may be nailed to the bottom of the carriage and to the joists to help tie the carriage to the floor frame. Fitting the carriage tightly against the header joist also helps to support the carriage.

To fit a 2 × 10 carriage to 2 × 10 joists, a 2 × 2 is nailed to the underside of the carriage and to the joist. Added support is obtained by placing a 2 × 2 block against the 2 × 2 nailed to the carriage and nailing it to the joist. To provide greater support, 2 × 4 posts resting on the floor can be nailed to the sides of the carriage.

Dadoed Stringers

A dadoed stringer may be made from a 2 × 6, 2 × 8, or larger lumber sizes. The dadoed stringer may be used when riser boards are not necessary, and a simple stair is desired. The treads in this type of stair are supported on the dadoes cut into the stringer. See Figure 6-8.

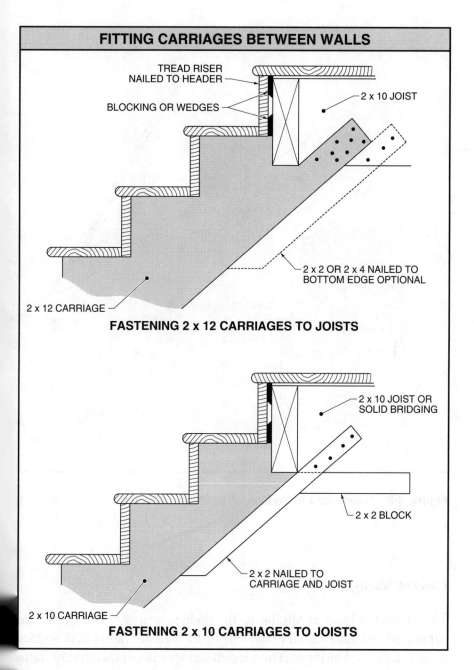

Figure 6-7. The carriages must be securely fastened in place.

100 STAIR LAYOUT

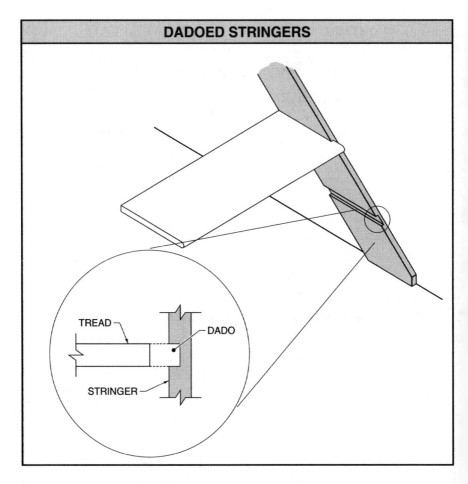

Figure 6-8. Treads rest in dadoes of a dadoed stringer.

Cleated Stringers

The cleated stringer is similar to the dadoed stringer. The treads are supported on cleats nailed to the side of the stringer rather than dadoes as in the dadoed stringer. The cleated stringer is comparatively inexpensive. See Figure 6-9.

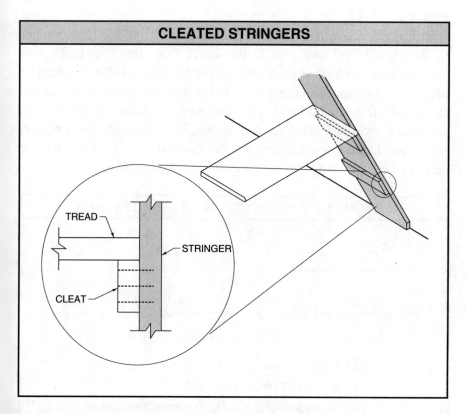

Figure 6-9. Treads rest on cleats of a cleated stringer.

Layout of Dadoed and Cleated Stringers. The layout procedure for these two types of stringers is very similar. To lay out a dadoed stringer, stair gauges are clamped to the framing square so that when the square is laid on the stringer, the unit rise and unit run dimensions fall on the edge of the stock. The stringer material is placed on sawhorses or a workbench of convenient height, and the location of the treads is marked in a manner similar to that of marking a stair carriage. The main difference is that the risers are not completely marked out. The rise is set on the tongue of the square and the run on the body of the square.

102 STAIR LAYOUT

It is good practice to start at the lower (left) end of the stringer by drawing a line to represent the lower floor line. Next, the first riser is indicated with a short line on the rise side of the square. The square is then slipped to the next position, and the first tread is marked along with the second riser. This procedure is followed until all of the treads are marked. After the layout is completed, the stringers are either dadoed or cleats are nailed along the layout lines for the purpose of supporting the stair treads. See Figure 6-10.

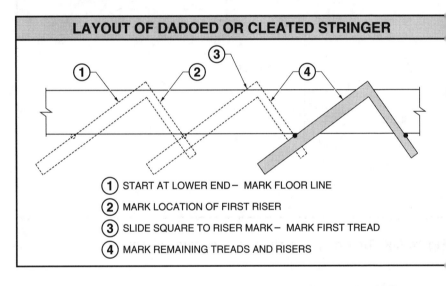

Figure 6-10. The layout of dadoed and cleated stringers is similar.

Housed Stringers

The housed stringer is one of the most commonly used stringers in residential construction. This stringer has a very good finished appearance and provides a "mop board" along the slope of the stair thereby making the stairs easy to clean. Housed stringers are usually made from 1 × 10 or 1 × 12 boards. The treads and risers are fitted into housed openings and held in place by wedges which are glued and driven under the treads and behind the risers. See Figure 6-11.

Stringers 103

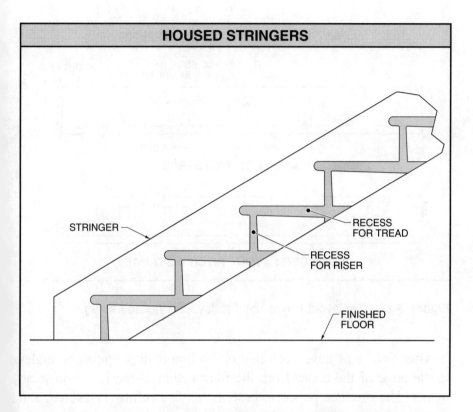

Figure 6-11. The housed stringer has a very good finished appearance.

Layout of a Housed Stringer. A router and stair-routing template are usually used to cut the recesses for the risers and treads. To lay out a housed stringer, draw a layout line $1\frac{1}{2}''$ to $2''$ from the bottom edge of a 1 × 12 stringer ($\frac{3}{4}''$ on a 1 × 10) for a distance of approximately 4'-0".

When using a routing template it is necessary to lay out only two steps using the unit rise and unit run. These steps are marked off using the layout line as a base rather than the edge of the board. See Figure 6-12. This layout line is used instead of the edge of the board because space must be allowed for the wedges which support the treads and risers.

104 STAIR LAYOUT

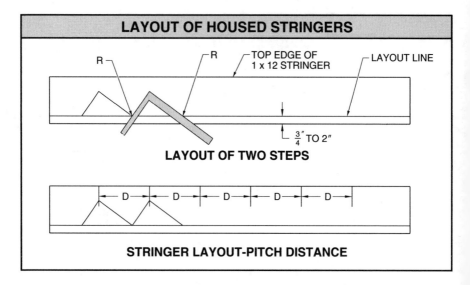

Figure 6-12. The layout line is used to layout a housed stringer.

After two steps have been laid out, a line is drawn at right angles to the edge of the board from the intersection of the riser and tread marks. The distance between these marks is carefully measured, and then equal distances are marked along the entire length of the stringer.

The router template is laid over one of the step outlines and adjusted so that when it is clamped in place, the riser face guide and the tread face guide are parallel to the riser and tread marks. After the template is clamped on the stringer, the index pointer on the template is aligned with the line drawn from the intersection of the riser and tread marks. The router is then set for the proper depth and the riser and tread space is housed out. To house out the remaining steps, the template is loosened and moved to the next index mark.

When making housed stringers, care must be taken to make a right-hand and a left-hand stringer. It is good practice to lay out the second stringer by laying it along side the first stringer and marking the top end, which will be at the lower end of the first stringer. When laid out in this manner, the framing square can be used to mark both stringers at the same time. To house out the second stringer, the router template is turned over (upside down) and clamped to the stringer in a manner similar to that used on the first stringer. See Figure 6-13.

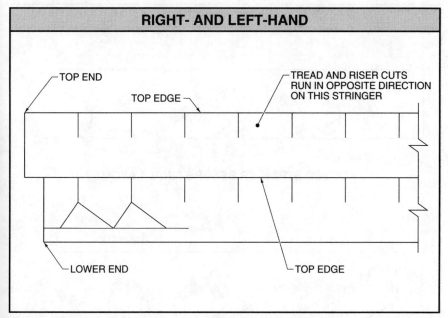

Figure 6-13. A right-hand and left-hand stringer are required for housed stringers.

Installing Housed Stringers. Housed stringers are generally delivered to the job with the ends running long. See Figure 6-14. Before they can be installed, the lower end must be cut to fit the floor and maintain a uniform unit rise. The upper end must be cut to fit around the floor framework.

On the lower end, a level line is drawn to represent the floor line at the bottom riser. Care is taken to maintain the unit rise of the stair from the top of the tread recess to the level line. The lower end must also be cut to fit into a baseboard, or in some cases simply cut vertically to provide a finished end. When the stringer must fit into a molded baseboard, it is cut so that the height of the stringer measured vertically from the level line to the top edge of the stringer is equal to the height of the baseboard. When the lower end of the stringer fits into a base of equal thickness, it is often mitered.

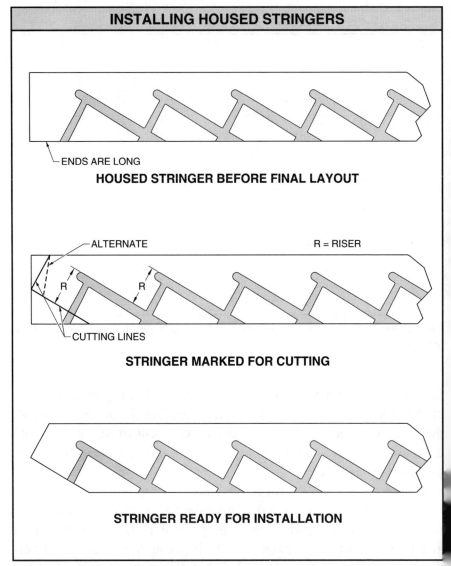

Figure 6-14. The ends of housed stringers run long when delivered to the job site.

The upper end of the stringer is laid out to fit around the floor construction. It is also laid out in such a way that the level cut above the starting tread or nosing is at base height. See Figure 6-15.

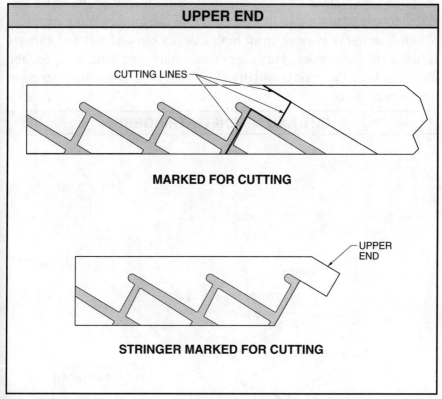

Figure 6-15. The upper end of the housed stringer is laid out to fit around floor construction.

Housed stringers are generally installed between walls and are fastened by nailing into the wall studs. Finish nails or casing nails are generally used for this purpose. However, common nails may be placed below the treads and behind the risers.

Cut and Mitered Stringers

The cut and mitered stringer is used on open stairways. This type of stringer may be cut from a 1 × 10, or larger, board. The treads are supported on square cuts and the risers are mitered to present a finished appearance without any end grain of the lumber being exposed. When a stairway is open from both sides, a cut and mitered stringer is used on both sides. However, when only one side is open, the closed side of the stair is usually supported on a housed-out stringer. See Figure 6-16.

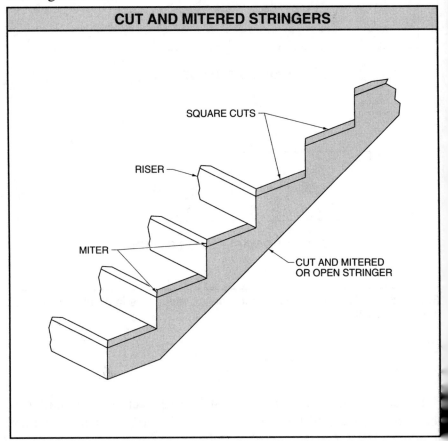

Figure 6-16. The cut and mitered stringer is used on open stairways.

Layout of Mitered Stringers. The layout of mitered stringers is similar to the layout of the cut-out carriage. With the cut-out carriage, the layout lines represent the back of the riser and the underside of the tread. When laying out a mitered stringer, the layout lines represent the face of the risers and the underside of the treads.

To lay out a mitered stringer, the outline of the risers and treads are laid out as for a carriage. After this is accomplished, the mitered risers are laid out by marking a second riser line parallel to the first at a distance equal to the stringer thickness. Making the layout in this manner results in the riser cut being made at exactly 45° with the face of the stringer. When power equipment is used to make the riser cut, the second layout is seldom required. See Figure 6-17.

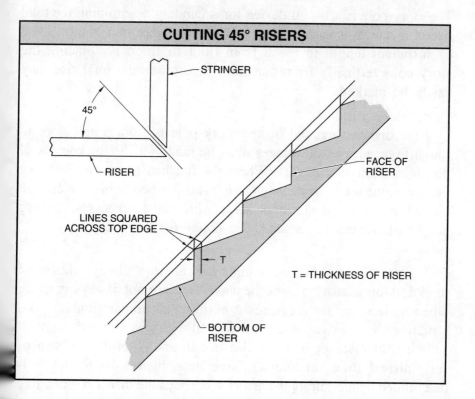

Figure 6-17. The riser is cut at 45° to the face of the stringer.

The installation of mitered stringers follows the same general rules applied to installing cut-out carriages and housed stringers. However, with the cut-out carriage the riser cut represents the back of the riser, while with the mitered stringer the front of the cut represents the riser face. Therefore, when these two types are being used in a stairway, care must be taken to properly align the stringers.

STORY POLE LAYOUT

The story pole is a useful device for accurately determining the total rise of a stair. It is usually made of a straight piece of 1 × 2 or 2 × 2 of sufficient length to reach from floor to floor. By placing the story pole vertically from one floor to another, the total rise may easily be marked.

If the distance marked on the story pole is from rough floor to rough floor, proper allowances must be made for the thicknesses of the finished floor materials. Where the finished floors are the same thickness, the total rise between rough and finished floors is identical, but when the finished floors are of different thicknesses, making proper allowances is essential. See Figure 6-18.

Because finished flooring materials are not always the same thickness on each floor, and because they are not always in place when a stair builder measures a building in preparation for constructing the stairway, the total rise should always be thought of as being the distance between finished floors. After the allowances for finished floor thicknesses have been made, the total rise is determined by measuring the distance between the marks representing the top of the finished floors. See Figure 6-19.

Stringers 111

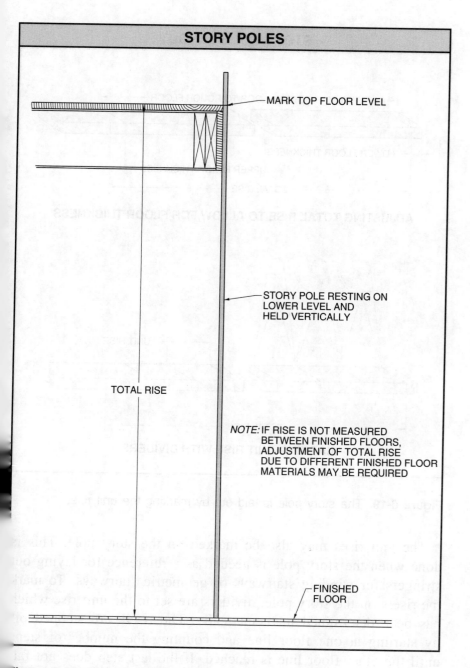

Figure 6-18. The story pole is a straight 1 × 2 or 2 × 2.

112 STAIR LAYOUT

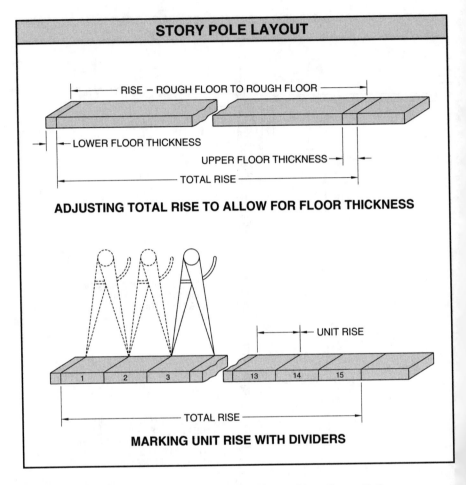

Figure 6-19. The story pole is laid out by marking the unit rise.

The unit rises may also be marked on the story pole. This is done when the story pole is needed as a reference for laying out stringers for winding stairways or geometric stairways. To mark the risers on the story pole, dividers are set to the unit rise which has been determined mathematically. The risers are stepped off by starting at one floor line and counting the number of steps until the other floor line is reached. If the last step does not fall exactly on the floor line, the dividers are readjusted, increasing slightly if the last step falls below the floor line or decreasing the rise slightly if the last step falls above the floor line. The story pole is then stepped off again. When the proper setting has been reached, the risers are marked on the story pole with a pencil and combination square and the risers are numbered.

PITCH DISTANCE AND MATHEMATICAL LENGTH OF STRINGER

The *pitch distance* of a step is the diagonal of the unit rise and unit run. It is found by measuring the distance between the intersection of two sets of risers and treads or by measuring the diagonal distance between the unit rise and unit run. See Figure 6-20.

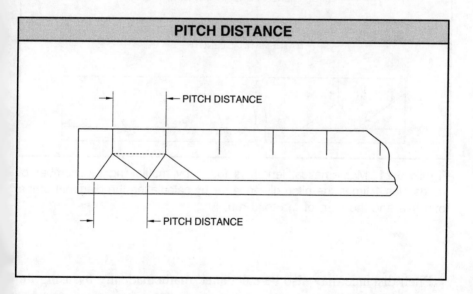

Figure 6-20. Pitch distance is the diagonal of the unit rise and unit run.

Mathematical length is determined after the pitch distance (hypotenuse) of a stair is found. *Mathematical length* is the number of treads plus one times the pitch distance. See Figure 6-21.

114 STAIR LAYOUT

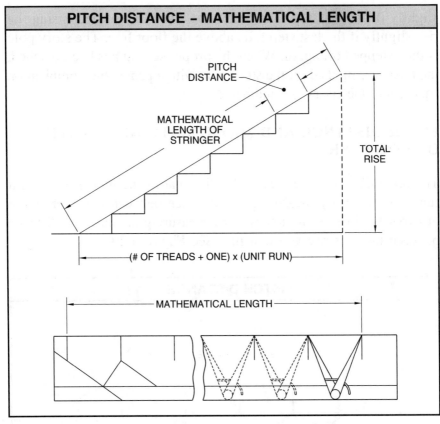

Figure 6-21. Mathematical length is found by multiplying the number of treads plus 1 times the pitch distance, or by calculating the diagonal of the total rise and the sum of the total run.

Pitch distance may also be calculated mathematically by using the Pythagorean Theorem. The Pythagorean Theorem states that the square of the hypotenuse of a right triangle is equal to the sum of the squares of the other two sides. The *hypotenuse* is the side of a right triangle opposite the right angle. Because a right triangle has a 3-4-5 relationship, it is often used in laying out right angles and checking corners for squareness. See Figure 6-22.

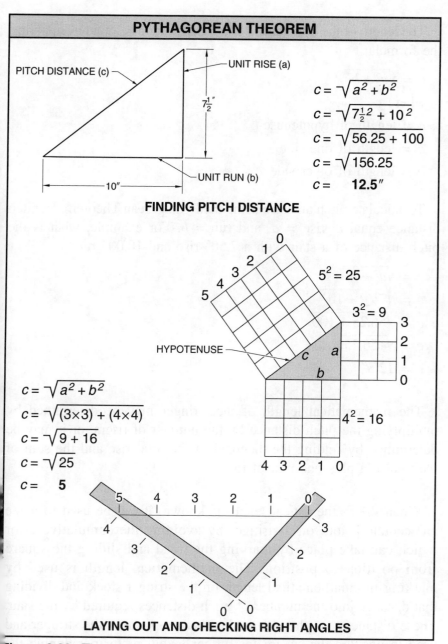

Figure 6-22. The square of the hypotenuse of a right triangle is equal to the sum of the squares of the other two sides.

116 STAIR LAYOUT

The length of the hypotenuse of a right triangle is found by applying the formula:

$$c = \sqrt{a^2 + b^2}$$

where

c = length of hypotenuse
a = length of one side
b = length of other side

To calculate pitch distance using the Pythagorean Theorem, let pitch distance equal c, rise = a, and run = b. For example, what is the pitch distance of a stair with a 7.50″ rise and 10.00″ run?

$$c = \sqrt{a^2 + b^2}$$
$$c = \sqrt{7\tfrac{1}{2}^2 + 10^2}$$
$$c = \sqrt{56.25 + 100}$$
$$c = \sqrt{156.25}$$
$$c = \mathbf{12.5″}$$

The mathematical length of the stringer may be determined by multiplying the pitch distance by the number of risers, or it may be determined by finding the diagonal of the total rise and the sum of the total run plus one unit of run.

When the mathematical length is known, it can be used to make an accurate layout on a stringer by avoiding the cumulative error which can take place by marking the tread and sliding the square from position to position. The mathematical length is used by marking the mathematical length on the stringer stock and dividing that distance into the number of pitch distances required by the stair. These distances can then be marked part way across the stringer and used to position the square for further layout or for positioning the router template.

Stringer Layout Steps

The basic steps followed in laying out the various stringers are:
1. Determine the total rise.
2. Determine the length of stairwell.
3. Determine the thickness of upper construction.
4. Determine the required headroom distance.
5. Determine the number of risers.
6. Determine the unit rise and unit run.
7. Determine the number of risers and treads in the open.
8. Check the unit run with the number of treads in the open to see if headroom is maintained.
9. Mark off the required number of steps of pitch distances.
10. Cut out the stringer or carriage and install to meet job conditions.

TRADE TEST 6
STRINGERS

Date _____ Name _____

_____ 1. The _____ carriage is one of the most common types of stringers.

_____ 2. Stairs are often laid out with the aid of _____ clamped to the square.

_____ 3. The _____ carriage consists of a 2 × 4 or 2 × 6 member with blocks fastened to it.

_____ 4. When laying out stair carriages, the _____ end of the stringer is on the layout person's _____.

    A. upper; right   C. lower; right
    B. upper; left   D. lower; left

_____ 5. Allowing space for the tread thickness at the bottom step of a carriage is known as "_____ the carriage."

    A. lowering   C. dropping
    B. raising   D. neither A, B, nor C

_____ 6. The _____ stringer provides a mop board along the slope of the stair.

_____ 7. When making _____ stringers, left- and right-hand stringers are required.

    A. dadoed   C. housed
    B. cleated   D. A, B, and C

120 STAIR LAYOUT

T   F   8. Housed stringers are generally delivered to the job with the ends running long.

T   F   9. The 2 × 12 is the most common size of lumber used for cut-out carriages in residential work.

_____ 10. A(n) _____ is a strip of lumber on which the total rise of the stair is marked.

_____ 11. The _____ of a step is the diagonal of the unit rise and unit run.

_____ 12. The Pythagorean Theorem is _____.

    A. $c = \sqrt{a+b}$ C. $c = \sqrt{a^2+b^2}$
    B. $c^2 = \sqrt{a+b}$ D. neither A, B, nor C

_____ 13. The _____ is the side of a right triangle opposite the right angle.

_____ 14. A right triangle has a _____ relationship.

    A. 1-2-3   C. 3-4-5
    B. 2-3-4   D. neither A, B, nor C

_____ 15. The unit rise and _____ must be determined before stair carriages are laid out.

T   F   16. Housed stringers are usually made from 1 × 10 or 1 × 12 boards.

_____ 17. A(n) _____ and stair-routing template are usually used to cut the recesses for risers and treads in housed stringers.

_____ 18. The cut and mitered stringer is used on _____ stairways.

_____ 19. Total _____ is the distance from finished floor to finished floor.

_____ 20. A right triangle may be used in _____.
   A. laying out right angles   C. both A and B
   B. checking corners for   D. neither A nor B
      squareness

_____ 21. A(n) _____ carriage is shown at A.

_____ 22. A(n) _____ carriage is shown at B.

_____ 23. A(n) _____ stringer is shown at C.

_____ 24. A(n) _____ stringer is shown at D.

_____ 25. A(n) _____ stringer is shown at E.

_____ 26. A(n) _____ stringer is shown at F.

122 STAIR LAYOUT

_____ 27. The pitch distance of the stringer at G is _____".

_____ 28. The mathematical length of the stringer at G is _____.

_____ 29. The pitch distance of the stringer at H is _____".

_____ 30. The mathematical length of the stringer at H is _____.

_____ 31. The pitch distance of the stringer at I is _____".

_____ 32. The mathematical length of the stringer at I is _____.

_____ 33. The pitch distance of the stringer at J is _____".

_____ 34. The mathematical length of the stringer at J is _____.

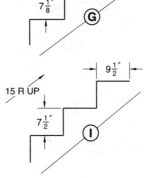

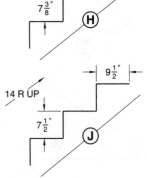

7 WINDERS

The layout of winding stairs and geometric stairs involves the use of the same general rules as layout for a straight stair. However, care must be taken to avoid narrow winding treads at the converging end. All treads should be the same width along the line of travel in winder and geometric stairs. Stringer layout requires a full-size plan view, story pole, and a large working surface.

WINDING STAIRS

Winding stairs are often used when there is insufficient space for a stairway with landings. These stairs save space because the space normally occupied by landings is used by winding treads. Winding stairs are dangerous, especially to small children, unless precautions are taken to avoid the very narrow width of treads at the converging end.

Winding stairs can be made reasonably safe by designing them with all treads of equal width along the line of travel and designing all winding treads of sufficient width at the converging end. The line of travel is usually placed 14" to 18" from the stringer on the converging side of the stair treads.

When the quarter-turn winder is used in place of the quarter-turn stair, it allows the designer to put at least two risers and three treads in the space occupied by the landing in the quarter-turn stair. The half-turn winder usually allows the use of five or six winding treads in place of the landing. Winding stairs should be provided with sufficient handrails.

Quarter-turn Winders

The layout of the plan view for the quarter-turn winder is made after the total rise, the number of risers, and the amount of space available for the stair is determined. The unit run is determined in part by the total available run for the quarter-turn winder. To determine the total run available, the two straight runs are added to the run included in the quarter-turn. The run of the quarter-turn is actually one-quarter of the circumference of a circle whose radius is equal to the sum of the radius of the well and line of travel distance. See Figure 7-1.

After determining the total run available and making the necessary allowances for riser thicknesses, the available run is divided by the number of treads to obtain the unit run. The unit run is then stepped off along the line of travel on the plan view to obtain the preliminary plan view layout. *Note:* The run around the quarter-turn was determined to be on the circumference of a circle, but it is stepped off as a chord of the circle. Therefore, the unit run of each tread around the turn must be decreased slightly (approximately $\frac{1}{16}''$) to allow for the errors in stepping off. This allowance is not reflected in Figure 7-1, but it would have to be made on an actual layout. A *chord* is a line from circumference to circumference of a circle not through the centerpoint.

The initial plan layout shows the winding treads to be very narrow at the converging end. As narrow treads at this point are a safety hazard, they must be increased in width. The increase in tread width is accomplished by pivoting the riser line at the line of travel in such a manner as to increase the width of the treads at the inner end. See Figure 7-2.

After the plan view has been completed, it is good practice to check the headroom and the length of the wellhole. This is done in the same manner as for a straight or quarter-turn stair with a landing.

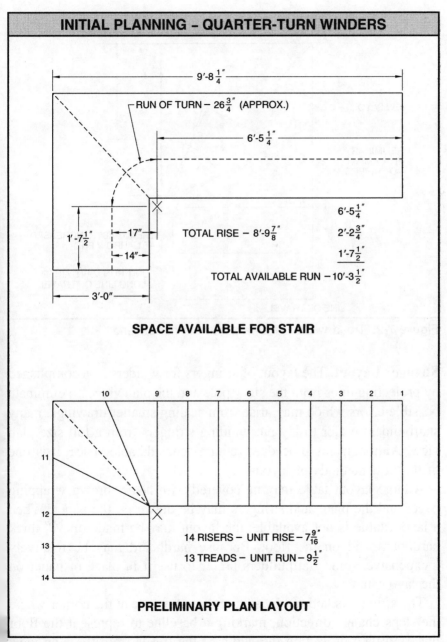

Figure 7-1. The plan view is developed after the total rise, number of risers, and amount of space is determined.

126 STAIR LAYOUT

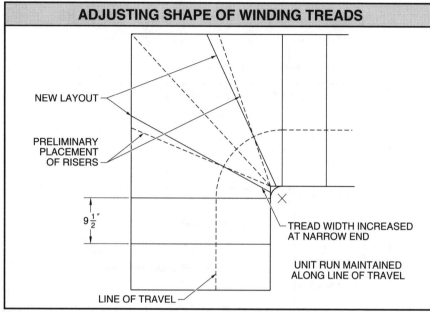

Figure 7-2. Tread width is increased at the narrow end.

Stringer Layout. The layout of stringers for winders is accomplished by projecting lines from the story pole and the plan view. To eliminate possible errors which may arise from scaling smaller drawings, many stairbuilders prefer to lay out winding stringers from a full-size plan view. Although this procedure requires considerable space, it is one of the best methods of layout.

A large layout table may be covered with heavy brown wrapping paper and the plans and stringers may be drawn on the paper. When a layout table is not available, the layout may be made on $\frac{1}{8}''$ thick hardboard laid on the floor. Because hardboard may be relatively inexpensive, some stairbuilders prefer to use it in place of paper on the layout table.

The stringer is laid out by setting the story pole at the corner where the steps change direction, marking a baseline to represent the floor line, and then projecting the width of the treads from the plan view and the riser heights from the story pole. After all the lines have been projected, the actual risers and treads are darkened in and the shape of the stringer becomes apparent. See Figure 7-3.

The width of the stringer board varies and becomes a matter of trial and error in order to get the best shape for the stringer. It is recommended that the stringer be made horizontal at the point where the stair changes direction so that moldings running on the top edge of the stringers may be easily matched.

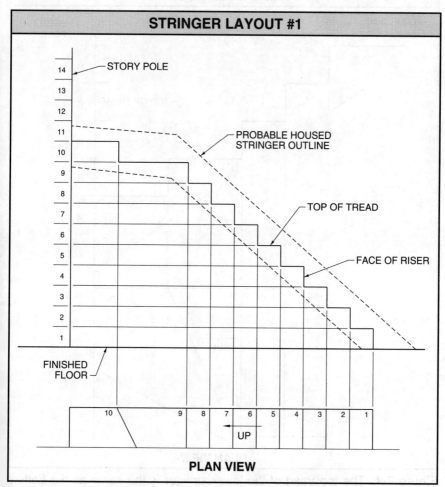

Figure 7-3. The story pole is used to lay out the stringer.

128 STAIR LAYOUT

The second half of the outside stringer is made by projecting the tread widths from plan view in a similar manner as for the lower portion of the stringer. Care must be taken to be sure that the story pole is aligned properly to maintain the proper layout. The top tread of the lower stringer and the bottom tread of the top stringer are one and the same. See Figure 7-4.

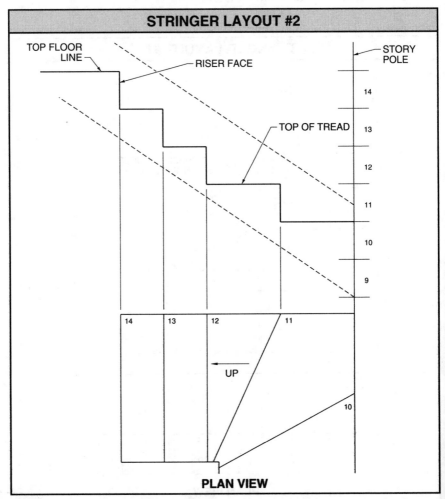

Figure 7-4. The top tread of the lower stringer is the same as the bottom tread of the top stringer.

The inside stringers are laid out in a manner similar to that for the outside stringers. Due to the manner in which the stair is projected, the work may seem upside down. See Figures 7-5 and 7-6.

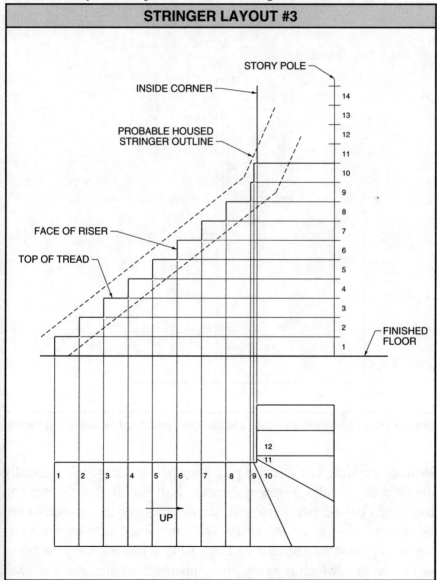

Figure 7-5. The stringers are laid out from the story pole.

130 STAIR LAYOUT

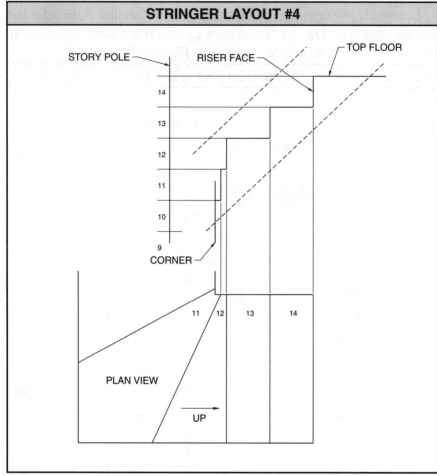

Figure 7-6. The inside stringers are laid out similar to the outside stringers.

Winder Layout. The full-size plan view of a winding stair is usually drawn with the sides represented by the wall line. Before the winding treads can be accurately laid out, the thickness of the stringers must be drawn. Also, since the treads are represented without a nosing, the nosing must also be drawn. The nosing is represented by a heavy or broken line, which is easily distinguished from the riser lines. See Figure 7-7.

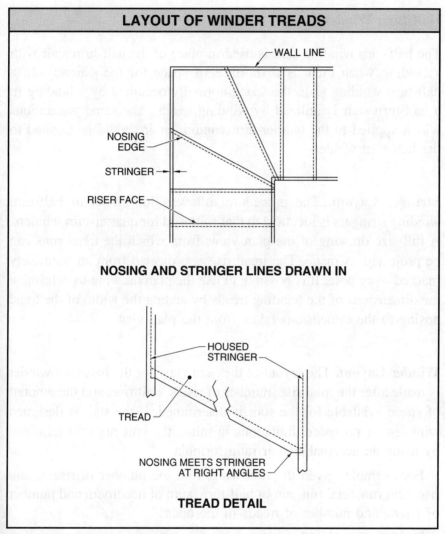

Figure 7-7. The full-size plan view of a winding stair is usually drawn with the sides represented by the wall line.

After this preliminary layout work, the dimensions of the winding treads may be easily measured. To make fitting treads to a housed stringer easier, the tread nosing may be returned at right angles to the stringer.

Half-turn Winders

The half-turn winding stair is used in place of the half-turn stair with a landing when there is a shortage of space for the stairway. In a half-turn winding stair, the space normally occupied by a landing in a half-turn stair is utilized by winding treads. The same precautions which applied to the quarter-turn winding stair should be applied to the half-turn winder.

Stringer Layout. The procedure followed for layout of half-turn winding stringers is identical to that followed for quarter-turn winders. A full-size drawing of the plan view, from which the tread runs can be projected, is made. The tread rise is projected from an accurately marked story pole. It is possible to use the plan drawing to determine the dimensions of the winding treads by adding the width of the tread nosing to the dimensions taken from the plan view.

Winder Layout. The layout of the plan view for the half-turn winder is made after the total rise, number of risers, unit rise, and the amount of space available for the stair is determined. If the stair is designed with few or no space limitations in mind, the unit run is determined by using an acceptable stair ratio formula.

For example, given the following, find the number of risers, unit rise, unit run, total run, run of half-turn, sum of headroom and number of risers, and number of treads in the open.

Total Rise = 9'-10"
Upper Construction = 1'-4"
Desired Headroom = 7'-0"

1. Determine number of risers.
   $118'' \div 7'' = \mathbf{16}$
2. Determine unit rise.
   $118'' \div 16 = \mathbf{7\frac{3}{8}''}$

3. Determine unit run. Formula 1: *Unit Rise plus Unit Run equals 17" to 18"*. Use 17"
   $17" - 7\frac{3}{8}" = \mathbf{9\frac{5}{8}"}$
4. Determine total run. Unit Run × Number of Treads.
   Total run = $9\frac{5}{8}" \times 15 = \mathbf{144\frac{3}{8}"}$
5. Determine run of half-turn.
   $$\pi \times \frac{(Diameter\ of\ turn)}{2}$$
   $3.14 \times \frac{36}{2} = \mathbf{56.5"}$
6. Determine sum of headroom and upper construction.
   $7'\text{-}0" + 1'\text{-}4" = \mathbf{8'\text{-}4"}$
7. Determine number of treads in the open.
   $8'\text{-}4" \div 7\frac{3}{8}" = 13.6 = \mathbf{14}$

When space is limited, the unit run is determined in part by the total run available for the stair along with the number of treads used in the half-turn and the stair ratio formulas. The total run of the half-turn winding stair is determined by adding the run of the half-turn, which is one-half the circumference of a circle with a radius equal to the sum of the line of travel distance plus the radius of the well, and the two straight runs.

This total run is divided by the number of treads to determine the unit run. The resulting unit run is checked with the unit rise in one of the stair ratio formulas. If the combination fits the stair ratio formula, the plan view is laid out. If the combination does not fit the formula, either the unit rise, unit run, or both are adjusted to fit the formula.

The plan view of a half-turn winder stair shows the amount of run used in the turn and how it is laid out along the line of travel so that all treads are of equal width along the line of travel. See Figure 7-8.

In order to maintain a minimum headroom of 7'-0" for this stair, the stairwell must be large enough to accommodate at least 14 treads in the open. The number of treads in the open is calculated in the same manner as for a straight stair.

134 STAIR LAYOUT

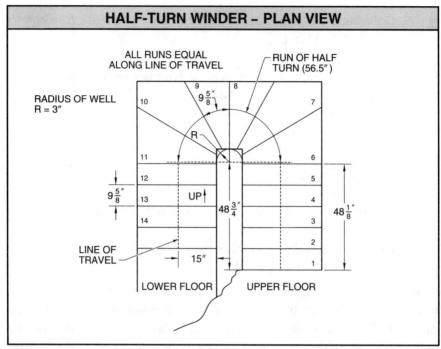

Figure 7-8. The plan view of a half-turn winder stair shows the amount of run used in the turn.

Geometric Stairways

The layout principles of straight stairways may be applied to geometric stairs. In the layout of geometric stairs the total rise, number of risers, and unit rise are determined in the usual manner. The unit run is then obtained by applying one of the stair ratio formulas, or by dividing the total run by the number of treads.

With this information, the plan view is developed. In laying out treads in plan view, care must be taken to see that all tread runs are equal along the line of travel.

The building of geometric stairs usually requires the construction of forms which coincide with the limits of the plan view and are high enough to accommodate the total rise. The layout of the stair can then be made on the curved form.

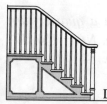

TRADE TEST 7
WINDERS

Date _____ Name _____

T F   1. Winding stairs are often used when there is insufficient space for a stairway with landings.

_____ 2. The line of travel for winding stairs is usually placed _____ from the stringer on the coverging side of the stair treads.

   A. 12" to 16"   C. 14" to 16"
   B. 12" to 18"   D. 14" to 18"

_____ 3. A(n) _____ is a line from circumference to circumference of a circle not through the centerpoint.

_____ 4. The full-size plan view of a winding stair is usually drawn with the sides represented by the _____ line.

   A. finished floor   C. either A or B
   B. wall   D. neither A, B, nor C

T F   5. The procedure followed for layout of half-turn winding stringers is identical to that followed for quarter-turn winders.

_____ 6. The layout of stringers for quarter-turn winders is accomplished by projecting lines from the _____ and the plan view.

136 STAIR LAYOUT

Use Stairway A for Problems 7 through 12. Design a quarter-turn winding stair to use a minimum amount of plan area. Note: Use 17" minus unit rise equals unit run. The line of travel is 14".

_____ 7. The number of risers is _____.

_____ 8. The unit rise is _____".

_____ 9. The unit run is _____".

_____ 10. The number of risers and treads in the open is _____.

_____ 11. The run of the one-quarter turn is _____".

_____ 12. The layout of the stringer is made with a _____.

        A. scale drawing   C. story pole
        B. full size drawing   D. B and C

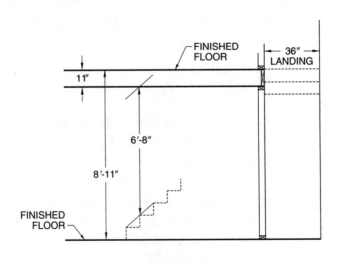

STAIRWAY A

Use Stairway B for Problems 13 through 18. Design a half-turn winding stair to use a minimum amount of plan area. Note: Use 17" minus unit rise equals unit run. The line of travel is 14".

_____ 13. The number of risers is _____.

_____ 14. The unit rise is _____".

_____ 15. The unit run is _____".

_____ 16. The number of risers and treads in the open is _____.

_____ 17. The run of the one-half turn is _____".

_____ 18. The plan view can be laid out after the _____ is determined.

        A. unit run   C. A and B
        B. number of risers   D. total rise
           in the open

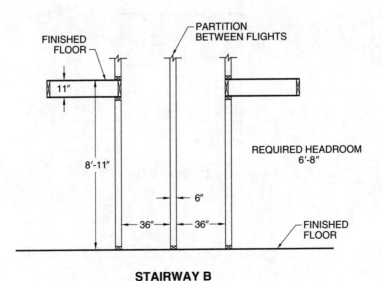

STAIRWAY B

19. Stairway C is a plan view of a(n) _____ winder. _____

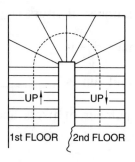

STAIRWAY C

20. Stairway D is a plan view of a(n) _____ winder. _____

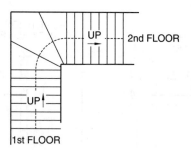

STAIRWAY D

8 STAIR SAFETY

In developing any stair design, it is necessary to follow building code requirements and to also design a stair which is easy and safe to use. Additionally, the cost of the stairway, the space it occupies, and the amount of traffic it carries must be considered. The stair must be easy to use by the majority of the people who use it.

SAFETY CONSIDERATIONS

In recent years, a number of studies have been done on the causes of accidents on stairways and injuries caused by those accidents. The results were inconclusive in nearly all of these studies.

However, it is apparent that stairs should be built in a manner which directs the user's attention to the stairway. Flooring patterns which cause the stair treads to blend in with the surrounding floor areas should be avoided. People are more likely to misstep if they are not aware of the exact location of the steps.

The relationship between the height (unit rise) and width (unit run) of a flight of stairs also is a factor in the safety of the stairs. Generally, keeping the unit rise between $6\frac{5}{8}''$ and $7\frac{3}{4}''$ and applying one of the established stair formulas results in a stair which is safe and easy to use.

At least one study concluded that the unit rise should not exceed $7''$ and tread run should be a minimum of $11''$. Some building codes have accepted this standard and others are considering it. There is no conclusive proof that this requirement results in a safer stair.

140 STAIR LAYOUT

If this requirement must be followed, it results in more space being occupied by the stair and less space available for other activity. To keep the usable space the same, the building must be enlarged. This results in added cost to the owner.

For example, in a comparison between a stair designed according to older codes which comply with accepted trade practices and a stair in which the unit rise does not exceed 7", an additional 4'-0" of run is required. This is for a typical dwelling with an 8'-0" ceiling and a total rise of 106⅞". Using standard rules of stair layout, this dwelling would have 14 risers at 7⅝". See Figure 8-1.

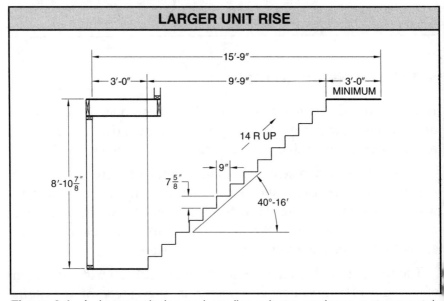

Figure 8-1. A larger unit rise and smaller unit run produces a steeper stair.

Total Rise = 106⅞"
Required Headroom = 80"
Upper Construction = 10⅞"
1. Determine number of risers.
   Total Rise ÷ 8
   106⅞" ÷ 8 = 13+ = **14**

2. Determine unit rise.
   Total Rise ÷ Number of Risers
   $106\frac{7}{8}" ÷ 14 = 7.63$ or $\mathbf{7\frac{5}{8}"}$
3. Unit run based on code minimum = **9"**
4. Check: Formula 2:
   $9" + 2(7\frac{5}{8}") = \mathbf{24\frac{1}{4}"}$
5. Determine number of risers in open.
   Headroom + Upper Construction ÷ Unit Rise
   $90\frac{7}{8}" ÷ 7\frac{5}{8}" = 11.9 = \mathbf{12}$
6. Determine length of stairwell.
   Unit Run × Number of Risers in Open
   $9 × 12 = \mathbf{108"}$
7. Determine total run.
   Unit Run × Number of Treads
   $9 × 13 = \mathbf{117"}$

If the stair must be built with a maximum unit rise of 7", the total rise is divided by 7" and rounded up to the next full number. See Figure 8-2.

Total Rise = $106\frac{7}{8}"$
Maximum Unit Rise = 7"
Required Headroom = 80"
Upper Construction = $10\frac{7}{8}"$

1. Determine number of risers.
   Total Rise ÷ 7"
   $106\frac{7}{8}" ÷ 7" = 15+ = \mathbf{16}$
2. Determine unit rise.
   Total Rise ÷ Number of Risers
   $106\frac{7}{8}" ÷ 16 = 6.68$ or $\mathbf{6\frac{11}{16}"}$
3. Unit run based on code minimum = **11"**
4. Check: Formula 2:
   $11" + 2(6\frac{11}{16}") = \mathbf{24\frac{3}{8}"}$
5. Determine number of risers in open.
   Headroom + Upper Construction ÷ Unit Rise
   $90\frac{7}{8}" ÷ 6\frac{11}{16}" = 13.6 = \mathbf{14}$

142 STAIR LAYOUT

6. Determine length of stairwell.
   Unit Run × Number of Risers in Open
   11 × 14 = **154″**
7. Determine total run.
   Unit Run × Number of Treads
   11″ × 15 = **165″**

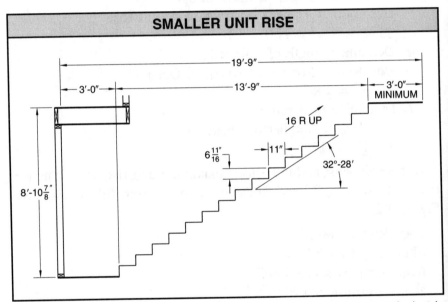

Figure 8-2. A smaller unit rise and larger unit run produces a stair that is not as steep.

Each of these stairways was designed using established stair design methods and applicable building codes. Each is probably as satisfactory as the other, but the second design uses considerably more floor space. Assuming a stairway 36″ wide, the first stair requires a minimum of 56.15 sq ft and the second stairway requires a minimum of 79.74 sq ft.

FINAL EXAM

Date _____ Name _____

_____ 1. The minimum preferred stair pitch at A is _____°.

_____ 2. The maximum preferred stair pitch at B is _____°.

_____ 3. All stairways serving an occupancy load of 49 people or less must be at least _____" wide.

_____ 4. Stairways for dwellings must have at least _____ handrail(s).

_____ 5. Handrails are located _____" to _____" above the nosing of the stairway treads.

        A. 28; 32   C. 30; 36
        B. 30; 34   D. 32; 36

T   F 6. Stairways with two or more risers must meet code standards.

144 STAIR LAYOUT

T   F   7. The sum of one tread and one riser, exclusive of the nosing, should not be more than 24″ nor less than 18″.

_____   8. The minimum length of a landing should be _____″.

_____   9. A(n) _____ is a vertical support for a rail.

_____   10. Total _____ is the vertical distance from finished floor to finished floor.

_____   11. The total run of a stairway with 16 risers and a unit run of 10″ is _____.

_____   12. The unit rise of a stairway with 13 risers and a total rise of 95⅞″ is _____″.

_____   13. The _____ is the line along which most people walk on a stairway.

_____   14. A(n) _____ is the space or opening in the floor through which the stair passes.

_____   15. Riser height variance in a flight of stairs should not exceed _____″.

_____   16. Distance between landings should not exceed _____ vertically.

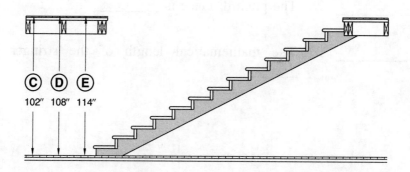

Use Formula 1 for Problems 17 through 25.

_____ **17.** The number of risers at C is _____.

_____ **18.** The unit rise at C is _____″.

_____ **19.** The unit run at C is _____″.

_____ **20.** The number of risers at D is _____.

_____ **21.** The unit rise at D is _____″.

_____ **22.** The unit run at D is _____″.

_____ **23.** The number of risers at E is _____.

_____ **24.** The unit rise at E is _____″.

_____ **25.** The unit run at E is _____″.

146 STAIR LAYOUT

Use Stringer F for Problems 26 and 27.

_____ 26. The pitch distance is _____".

_____ 27. The mathematical length of the stringer is _____.

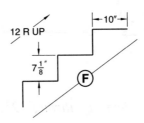

Use Stringer G for Problems 28 and 29.

_____ 28. The pitch distance is _____".

_____ 29. The mathematical length of the stringer is _____.

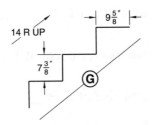

30. The length of the stairwell at H is _____.

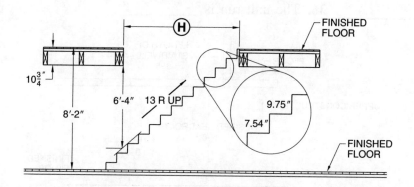

31. The length of the stairwell at I is _____.

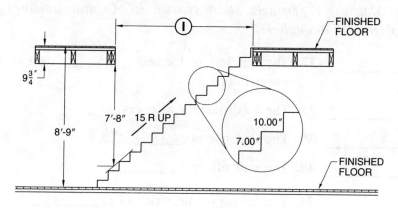

Use Stairway J for Problems 32 through 36.

_____ **32.** The number of risers is _____.

_____ **33.** The unit rise is _____".

_____ **34.** The sum of the headroom and upper construction is _____".

148 STAIR LAYOUT

_____ 35. The number of risers in the open is _____.

_____ 36. The unit run is _____".

```
                    FINISHED        LENGTH OF
                    FLOOR           STAIRWELL
                                    9'-5"

        UPPER CONSTRUCTION 11¾"

                        DESIRED HEADROOM
                            6'-8"         TOTAL RISE
                                          9'-0¾"
                                                        FINISHED
                                                        FLOOR
```

STAIRWAY J

Use Stairway K (Straight Stairway with 36" Central Landing) for Problems 37 through 41.

_____ 37. The landing is located _____" above the lower floor.

_____ 38. The total number of risers is _____.

_____ 39. The unit rise is _____".

_____ 40. The unit run is _____".

_____ 41. The length of the stairwell is _____.

```
                    FINISHED
                    FLOOR

        UPPER CONSTRUCTION 16"

                        DESIRED HEADROOM
                            7'-0"         TOTAL RISE
                                          16'-2"
                                                        FINISHED
                                                        FLOOR
```

STAIRWAY K

ANSWERS

Chapter 1 — Stair Basics

Trade Test 1   17
1. A
2. B
3. slope
4. 4
5. 3/8
6. 1½
7. plan
8. section
9. line of travel
10. tread
11. B
12. 30
13. 35
14. 8
15. 9
16. 7½
17. 10
18. run
19. riser
20. story pole
21. unit
22. total
23. B
24. C
25. D
26. T
27. F
28. 1¾
29. 44
30. 44; 88
31. 88
32. straight

Chapter 2 — Stair Design

Trade Test 2   33
1. D
2. T
3. T
4. Total rise
5. rise
6. unit run
7. C
8. B
9. A
10. sectional

11. **13**
12. **7.42**
13. **10.08**
14. **15**
15. **7.20**
16. **10.30**
17. **15**
18. **7.37**
19. **10.13**
20. **13**
21. **7.42**
22. **9.66**
23. **15**
24. **7.20**
25. **10.10**
26. **15**
27. **7.37**
28. **9.76**
29. **13**
30. **7.42**
31. **9.91**
32. **15**
33. **7.20**
34. **10.21**
35. **15**
36. **7.37**
37. **9.97**

Chapter 3
Stairwells

Trade Test 3   45

1. **B**
2. **C**
3. **D**
4. **T**
5. **T**
6. **B**
7. **C**
8. **B**
9. **B**
10. **T**
11. **10′-11⅛″**
12. **9′-4⅛″**
13. **12′-1″**
14. **13′-0″**
15. **15**
16. **7.33**
17. **10.17**
18. **12.50**
19. **10′-7⅛″**
20. **13**
21. **7.46**
22. **10.04**
23. **11.70**

24. 9′-9⁷⁄₁₆″
25. **16**
26. **7.25**
27. **10.25**
28. **12.60**
29. 10′-9⅛″
30. **17**
31. **7.47**
32. **10.03**
33. **12.80**
34. 10′-8⅜″
35. **18**
36. **7.56**
37. **9.94**
38. **12.60**
39. 10′-6″

Chapter 4
Headroom

Trade Test 4   **61**
1. T
2. D
3. 6′-4″
4. B
5. B
6. T

7. F
8. 7
9. 8
10. unit rise
11. 14
12. 7.77
13. 91.75
14. 11.90
15. 9.50
16. 14
17. 7.37
18. 91.75
19. 12.50
20. 9.28
21. 14
22. 7.41
23. 89.75
24. 12.10
25. 9.75
26. 14
27. 7.57
28. 89.75
29. 11.90
30. 10.34
31. 15
32. 7.34
33. 89.75
34. 12.20
35. 9.51

Chapter 5 — Landings

Trade Test 5   85

1. platform
2. C
3. D
4. B
5. A
6. 7
7. 8
8. 97
9. 26
10. 7.46
11. 10.04
12. 13′-5½″
13. 88
14. 24
15. 7.33
16. 10.17
17. 13′-9³⁄₁₆″
18. 14
19. 7.57
20. 9.43
21. 83.27
22. 16
23. 7.625
24. 9.375
25. 99.125
26. 10′-0″
27. 15
28. 7.47
29. 9.53
30. 89.64
31. 10′-1¾″

Chapter 6 — Stringers

Trade Test 6   119

1. cut-out
2. stair gauges
3. built-up
4. B
5. C
6. housed
7. D
8. T
9. F
10. story pole
11. pitch distance
12. C
13. hypotenuse
14. C
15. unit run
16. T

17. router
18. open
19. rise
20. C
21. cut-out
22. built-up
23. dadoed
24. cleated
25. housed
26. cut and mitered
27. 12.28
28. 12′-3⅜″
29. 12.125
30. 14′-1¾″
31. 12.10
32. 15′-1½″
33. 12.10
34. 14′-1⅜″

Chapter 7
Winders

Trade Test 7   135

1. T
2. D
3. chord
4. B
5. T

6. story pole
7. 14
8. 7.64
9. 9.36
10. 12
11. 22
12. D
13. 14
14. 7.64
15. 9.36
16. 12
17. 53⅜
18. C
19. half-turn
20. quarter-turn

Final Exam

Final Exam   143

1. 30
2. 35
3. 36
4. one
5. C
6. T
7. F
8. 36
9. baluster
10. rise

11. 12'-6"
12. 7⅜"
13. line of travel
14. wellhole
15. ⅜
16. 12'-0"
17. 13
18. 7.42
19. 10.08
20. 15
21. 7.20
22. 10.30
23. 15
24. 7.37
25. 10.13
26. 12.28

27. 14'-3⅜"
28. 12.125
29. 14'-1¾"
30. 9'-4⅛"
31. 12'-1"
32. 14
33. 7.77
34. 91.75
35. 11.90
36. 9.50
37. 97
38. 26
39. 7.46
40. 10.04
41. 13'-5½"

APPENDIX

ALPHABET OF LINES

NAME AND USE	CONVENTIONAL REPRESENTATION	EXAMPLE
OBJECT LINE Define shape. Outline and detail objects.	THICK	OBJECT LINE
HIDDEN LINE Show hidden features.	$\frac{1}{8}''$ (3 mm) THIN $\frac{1}{32}''$ (0.75 mm)	HIDDEN LINE
CENTER LINE Locate centerpoints of arcs and circles.	$\frac{1}{16}''$ (1.5 mm) THIN $\frac{1}{8}''$ (3 mm) $\frac{3}{4}''$ (18 mm) TO $1\frac{1}{2}''$ (36 mm)	CENTERLINE CENTERPOINT
DIMENSION LINE Show size or location. **EXTENSION LINE** Define size or location.	DIMENSION LINE — DIMENSION — 2'-6" THIN EXTENSION LINE	DIMENSION LINE $1\frac{3}{4}$ EXTENSION LINE
LEADER Call out specific features.	OPEN ARROWHEAD X THIN CLOSED ARROWHEAD — 3X	$1\frac{1}{2}$ DRILL LEADER
CUTTING PLANE Show internal features.	$\frac{1}{8}''$ (3 mm) THICK $\frac{1}{16}''$ (1.5 mm) A — A $\frac{3}{4}''$ (18 mm) TO $1\frac{1}{2}''$ (36 mm)	A — A LETTER IDENTIFIES SECTION VIEW CUTTING PLANE LINE
SECTION LINE Identify internal features.	$\frac{1}{16}''$ (1.5 mm) THIN	SECTION LINES
BREAK LINE Show long breaks. **BREAK LINE** Show short breaks.	$\frac{3}{4}''$ (18 mm) TO $1\frac{1}{2}''$ (36 mm) THIN FREEHAND THICK	LONG BREAK LINE SHORT BREAK LINE

156 STAIR LAYOUT

DECIMAL EQUIVALENTS

Fraction	Decimal	Fraction	Decimal
1/64	.015625	33/64	.515625
1/32	.03125	17/32	.53125
3/64	.046875	35/64	.546875
1/16	.0625	9/16	.5625
5/64	.078125	37/64	.578125
3/32	.09375	19/32	.59375
7/64	.109375	39/64	.609375
1/8	.125	5/8	.625
9/64	.140625	41/64	.640625
5/32	.15625	21/32	.65625
11/64	.171875	43/64	.671875
3/16	.1875	11/16	.6875
13/64	.203125	45/64	.703125
7/32	.21875	23/32	.71875
15/64	.234375	47/64	.734375
1/4	.250	3/4	.750
17/64	.265625	49/64	.765625
9/32	.28125	25/32	.78125
19/64	.296875	51/64	.796875
5/16	.3125	13/16	.8125
21/64	.328125	53/64	.828125
11/32	.34375	27/32	.84375
23/64	.359375	55/64	.859375
3/8	.375	7/8	.875
25/64	.390625	57/64	.890625
13/32	.40625	29/32	.90625
27/64	.421875	59/64	.921875
7/16	.4375	15/16	.9375
29/64	.453125	61/64	.953125
15/32	.46875	31/32	.96875
31/64	.484375	63/64	.984375
1/2	.500	1	1.000

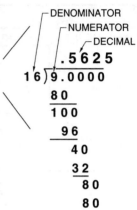

FORMULA 1					
UNIT RISE	UNIT RUN	DEGREE OF INCLINE (APPROX.)	UNIT RISE	UNIT RUN	DEGREE OF INCLINE (APPROX.)
6⅝"	10⅞"	31°-21'	7⅜"	10⅛"	36°-5'
6¹¹⁄₁₆"	10¹³⁄₁₆"	31°-44'	7⁷⁄₁₆"	10¹⁄₁₆"	36°-28'
6¾"	10¾"	32°-7'	7½"	10"	36°-52'
6¹³⁄₁₆"	10¹¹⁄₁₆"	32°-31'	7⁹⁄₁₆"	9¹⁵⁄₁₆"	37°-16'
6⅞"	10⅝"	32°-54'	7⅝"	9⅞"	37°-40'
6¹⁵⁄₁₆"	10⁹⁄₁₆"	33°-18'	7¹¹⁄₁₆"	9¹³⁄₁₆"	38°-5'
7"	10½"	33°-40'	7¾"	9¾"	38°-29'
7¹⁄₁₆"	10⁷⁄₁₆"	34°-5'	7¹³⁄₁₆"	9¹¹⁄₁₆"	38°-53'
7⅛"	10⅜"	34°-29'	7⅞"	9⅝"	39°-17'
7³⁄₁₆"	10⁵⁄₁₆"	34°-52'	7¹⁵⁄₁₆"	9⁹⁄₁₆"	39°-41'
7¼"	10¼"	35°-16'	8"	9½"	40°-5'
7⁵⁄₁₆"	10³⁄₁₆"	35°-40'			

Formula 1. *Unit Rise plus Unit Run equals 17" to 18"*.

Example. 6⅝" + 10⅞" = **17½"**

FORMULA 2					
UNIT RISE	UNIT RUN	DEGREE OF INCLINE (APPROX.)	UNIT RISE	UNIT RUN	DEGREE OF INCLINE (APPROX.)
6⅝"	11¼"	30°-29'	7⅜"	9¾"	37°-6'
6¹¹⁄₁₆"	11⅛"	31°	7⁷⁄₁₆"	9⅝"	37°-42'
6¾"	11"	31°-32'	7½"	9½"	38°-17'
6¹³⁄₁₆"	10⅞"	32°-4'	7⁹⁄₁₆"	9⅜"	38°-52'
6⅞"	10¾"	32°-36'	7⅝"	9¼"	39°-30'
6¹⁵⁄₁₆"	10⅝"	33°-8'	7¹¹⁄₁₆"	9⅛"	40°-7'
7"	10½"	33°-40'	7¾"	9"	40°-44'
7¹⁄₁₆"	10⅜"	34°-12'	7¹³⁄₁₆"	8⅞"	41°-21'
7⅛"	10¼"	34°-48'	7⅞"	8¾"	42°
7³⁄₁₆"	10⅛"	35°-21'	7¹⁵⁄₁₆"	8⅝"	42°-38'
7¼"	10"	35°-56'	8"	8½"	43°-15'
7⁵⁄₁₆"	9⅞"	36°-31'			

Formula 2. *Two Unit Rises plus Unit Run equals 24" to 25".*

Example. 2(6⅝") + 11¼" =
13¼" + 11¼" = **24½"**

FORMULA 3

UNIT RISE	UNIT RUN	DEGREE OF INCLINE (APPROX.)	UNIT RISE	UNIT RUN	DEGREE OF INCLINE (APPROX.)
6⅝"	11.09"	30°-51'	7⅜"	9.96"	36°-31'
6¹¹⁄₁₆"	10.99"	31°-19'	7⁷⁄₁₆"	9.88"	36°-58'
6¾"	10.89"	31°-47'	7½"	9.80"	37°-26'
6¹³⁄₁₆"	10.79"	32°-16'	7⁹⁄₁₆"	9.72"	37°-53'
6⅞"	10.69"	32°-45'	7⅝"	9.64"	38°-21'
6¹⁵⁄₁₆"	10.59"	33°-14'	7¹¹⁄₁₆"	9.56"	38°-49'
7"	10.50"	33°-40'	7¾"	9.48"	39°-16'
7¹⁄₁₆"	10.41"	34°-8'	7¹³⁄₁₆"	9.41"	39°-43'
7⅛"	10.32"	34°-37'	7⅞"	9.33"	40°-10'
7³⁄₁₆"	10.23"	35°-6'	7¹⁵⁄₁₆"	9.25"	40°-36'
7¼"	10.14"	35°-34'	8"	9.19"	41°-2'
7⁵⁄₁₆"	10.05"	36°-13'			

Formula 3. *Unit Rise multiplied by Unit Run equals 72 to 75.*

Example. $6\frac{5}{8}'' \times 11.09'' =$
$6.625'' \times 11.09'' = 73.47 = \mathbf{73\frac{1}{2}}$

STAIR LAYOUT

STANDARD LUMBER SIZES				
	THICKNESS		WIDTH	
TYPE	NOMINAL SIZE	ACTUAL SIZE	NOMINAL SIZE	ACTUAL SIZE
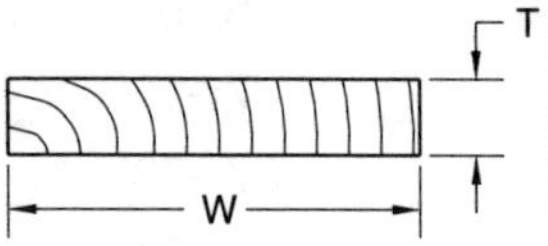 COMMON BOARDS	1"	¾"	2"	1½"
			4"	3½"
			6"	5½"
			8"	7¼"
			10"	9¼"
			12"	11¼"
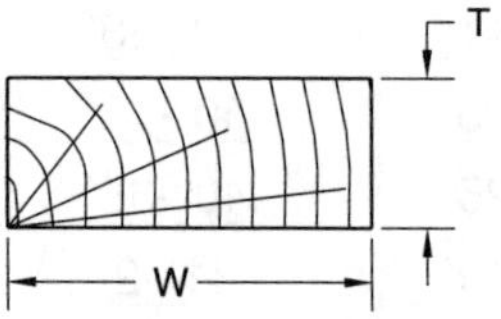 DIMENSION	2"	1½"	2"	1½"
			4"	3½"
			6"	5½"
			8"	7¼"
			10"	9¼"
			12"	11¼"
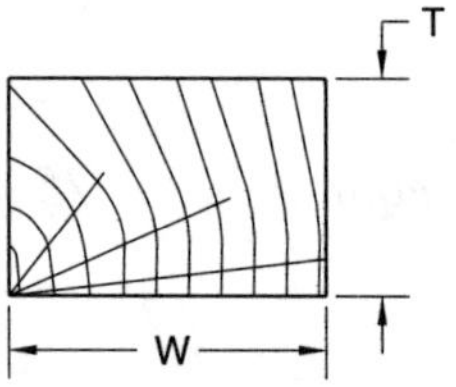 TIMBERS	4"	3½"	4"	3½"
	6"	5½"	6"	5½"
	8"	7½"	8"	7½"
			10"	9½"
	6"	5½"	6"	5½"
			8"	7½"
			10"	9½"
	8"	7½"	8"	7½"
			10"	9½"

GLOSSARY

angle: The number of degrees between two intersecting lines of a flat plane. For example, 45°.

angle

ANSI: American National Standards Institute.

architect: Person qualified and licensed to design and oversee construction of a building.

baluster: A vertical support for a rail. One of a series of slender vertical members of a balustrade; an upright support for the railing of a stairway. See *balustrade*.

balustrade: A row of balusters topped by a rail. See *baluster*.

baseboard: Molding placed at the base of a wall and fitted to the floor.

bevel: An angled cut from surface to surface of a board.

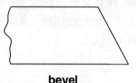
bevel

board foot: Unit of measure for lumber based on the volume of a piece 12″ square and 1″ thick. A board foot contains 144 cu in. (12″ × 12″ × 1″ = 144 cu in.)

building code: Regulations that establish required standards for the materials and methods of construction in a city, county, or state. Building codes are enforceable by law.

carriage: See *stringer*.

chamfer: An angled cut from a surface to an adjacent edge of a board.

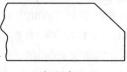
chamfer

chord: A line from circumference to circumference of a circle not through the centerpoint. See *circumference* and *circle*.

circle: A plane figure generated about a centerpoint. All circles contain 360°.

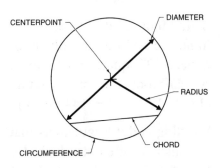

circle

circumference: The outside boundary of a circle. See *circle*.

clearance: The perpendicular distance from the stairway to the end of the stairwell. See *stairwell*.

conventional representation: A simplified way of representing building components on a print.

cutting plane: Line (identified by letters) that cuts through a part of a structure on a drawing. It refers to a separate elevation, sectional view, or detail drawing given for that area. See *elevation* and *sectional view*.

cutting plane

detail: Scaled plan, elevation, or sectional view drawn to a larger scale to show special features. See *elevation*, *plan*, and *sectional view*.

diameter: Distance from circumference to circumference of a circle through the centerpoint. See *circumference* and *circle*.

elevation: In printreading, the orthographic view of the exterior or interior walls. See *orthographic*.

flight of stairs: The series of steps leading from one floor to another or from a floor to a landing. See *landing*.

headroom: The vertical distance from the floor construction above the stair (at the end of the stairwell) to the slope of the stair. See *stairwell*.

horizontal: Level or parallel with the horizon.

horizontal

hypotenuse: The side of a right triangle opposite the right angle. See *angle*.

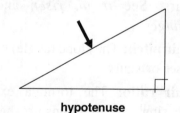

hypotenuse

landing: A horizontal platform separating two flights of stairs.

line of travel: The line along which most people walk on a stair.

mathematical length: The number of treads plus one times the pitch distance.

newel: Vertical post that supports a handrail at the top and bottom of a stairway. See *stairway*.

newel cap: Decorative top on a newel. See *newel*.

newel post: See *newel*.

nominal size: Descriptive size, not actual measured size. For example, 2″ × 4″ is the nominal size of a piece of wood actually measuring 1½″ × 3½″.

nosing: The portion of the tread which projects beyond the riser face. See *tread* and *riser*.

orthographic: Method of projecting planes at right angles. See *angle*.

plan view: A view looking down on the object.

Pythagorean Theorem: The square of the hypotenuse of a right triangle is equal to the sum of the squares of the other two sides. See *hypotenuse*.

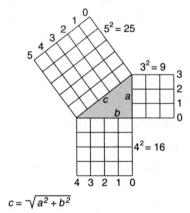

$c = \sqrt{a^2 + b^2}$

Pythagorean Theorem

radius: One-half the diameter of a circle. See *circle*.

rise: The vertical measure from the top of a tread to the top of the next higher tread. See *vertical* and *tread*.

riser: The vertical board between consecutive treads.

sectional view: A view made by passing a cutting plane through the object. May be either a plan or elevation view.

specifications: Written supplements to working drawings that give additional building information.

staircase: Entire assembly of stairs, landings, railings, and balusters.

stair gauge clamp: Small framing square accessory fastened in position to assist in laying out stairway stringers for treads and risers. Used in pairs. See *tread, riser,* and *stringer.*

stair pitch: The slope (angle) of a set of stairs.

stair ratio: The formula expressing the relationship between unit rise and unit run. See *unit rise* and *unit run.*

stairway: A whole set of steps; one or more flights of stairs.

stairwell (wellhole): The space or opening in the floor through which the stair passes.

starting newel: Post that supports the handrail at the bottom of a stairway. See *handrail* and *stairway*.

step: Unit consisting of one tread and one riser. See *tread* and *riser*.

story pole: A strip of lumber (usually a 1 × 2 or 2 × 2) on which the total rise of the stair is marked. See *rise*.

stringer (carriage): The part of the stair construction which is cut out to receive the risers and treads. The stringer (carriage) supports the steps. See *riser* and *tread*.

total number of (rises) risers: The number of unit rises in the stairway. See *unit rise*.

total number of treads: The number of unit runs in a given stairway. It is always one less than the total number of rises. See *unit run*.

total rise: The total vertical distance (height) of the stairs. It is measured from finished floor to finished floor.

total run: The total horizontal distance (length) of the stairs.

tread: The horizontal part of the stair which is walked on.

tread width: The width of the tread plus the width of the nosing. See *tread* and *nosing*.

unit rise: The height of each riser, which is the distance from the top of one step to the top of an adjacent step. See *riser*.

unit run (tread run): The width of each tread, not including the nosing. See *tread* and *nosing*.

upper construction: The total thickness of the floor and ceiling construction over the stairway. See *stairway*.

vertical: In a plumb or upright position.

vertical

INDEX

baluster, 15
balustrades, 15-*16*
building codes, 7
  handrail requirements, *9*
  stair standards, *8*

carriages, 91
  built-up, 92-*93*
  cut-out, *92, 97*
  layout, 93, *95*
  left long, *96*
circular and geometric stairs, *14*
cleated stringers, 100-*101*
  layout, 101-*102*
conventional representation, 1, 3
cut and mitered stringers, *108*
  layout, *109*

dadoed stringers, 98, *100*
  layout, 101-*102*
designing stairs, 5, 21-31, 51
dimensions, 3

enclosed and open stair construction, 14
baluster, 15
balustrades, 15-*16*
open stringer, *15*
existing conditions, 30

Formula 1, 21-*22*
Formula 2, 21, *23*
Formula 3, 21, *24*

gaining usable space, 41
geometric stairways, 134

half-turn stairs, 79-*80*, 132-*134*
  plan view, *81*
  sectional view, *83*
handrail requirements, 7, *9*
headroom, 4, 51, *52*-59
  designing stairs, 51

design steps, 59
determining, *56*
finding actual headroom, 57
housed stringers, *103*-107
 installing, *106*
 layout, *104*
 right- and left-hand, *105*
 upper end, *107*

incline of stairway, 10

landings, 4, 21, 67, *68*-83
 and risers, *70*
 half-turn stairs, 79-*80*
  plan view, *81*
  sectional view, *83*
 in straight stairs, 69
 located between floors, *72*
 quarter-turn stairs, 75-*76*
  plan view, *78*
  sectional view, *79*
 stairwell length, *74*
 types of, *12*
layout
 cleated stringers, 100-*101*
 dadoed stringers, 101-*102*
 half-turn winders, 132-*134*
 housed stringers, *103*-107
 mitered stringers, 108

quarter-turn winders, 75-76, 124-*131*
story pole, 110, *111-112*
winding treads, *126*
length of stairwell, 38-*39*
line of travel, 4
lumber for stairs, 91

mathematical length, 113-*114*
mitered stringers, 108
 layout, 109

nosing, 4
number of risers, 25, 27

open stringer, 15

pitch distance, *113*, *114*-116
plan view, 1
 half-turn winder, *134*
 quarter-turn layout, *131*
prints
 plan view, *2*

sectional view, 2
Pythagorean Theorem, 114, *115*-116

Q

quarter-turn stairs, 75-*76*, 124-131
   plan view, *78*
   sectional view, *79*
quarter-turn winders, 124-131
   initial planning for, *125*
   layout, *75-76*, 126-*131*

R

rise
   total, 5, 25
   unit, 5, 26
riser, 4
   cutting 45°, *109*
   total number of, 5
run
   total, 5, 29
   unit, 5, 26

S

safety, 139-142
sectional view, 1
stair design, 21, 43
   existing conditions, 30
   formulas, *22-24*
   number of risers, 25, *27*
   total rise, 25

total run, *29*
unit rise, 26
unit run, 26
stair gauges, 93-*94*
stair pitch, *6*
stair ratio, 4
   formulas, 21, *22-24*
stair safety, 139-142
stair standards, *8*
stair terminology, 3, *4*-6
stairway design, 5
   stair pitch, *6*
stairway types, 10
   circular and geometric stairs, *14*
   landings, *12*
   straight stairs, *11*
   winder stairs, *13*
stairwells, 37, *38*-43
   allowances, *41*
   calculation, 37
   gaining usable space, 41-*42*
   length, 38-*39*, 74
   steps, 43
story pole, 5, 73
   layout, 110, *111-112*
straight stair, *11*
stringers, 5, 91-117
   built-up carriages, 91-*93*
   cleated, 100
      layout, *101-102*
   cut and mitered, *108*
      layout, *109*
   cut-out carriages, 92, *97*
   dadoed, 98, *100*
      layout, 101, *102*

housed, *103*-107
  installing, *106*
  layout, *104*
  pitch distance, 113-*114*

terminology, 3, *4*-6
total number of (rises) risers, 5
total number of treads, 5
total rise, 5, 25
total run, 5, *29*
travel
  line of, *4*
tread, 5
tread width, 5

unit rise, 5, 26
unit run, 5, 26
upper construction, 5
usable space, *42*

wellhole, 5
winders, *13*, 123-134
  quarter-turn, 124-131
  half-turn, 132-*134*